on or before
iow.

Magnetic Resonance in Food Science
The Multivariate Challenge

Magnetic Resonance in Food Science
The Multivariate Challenge

Edited by

P.S. Belton
University of East Anglia, Norwich, UK

S.B. Engelsen
The Royal Veterinary and Agricultural University, Frederiksberg, Denmark

H.J. Jakobsen
University of Aarhus, Denmark

RS•C
advancing the chemical sciences

The proceedings of the 7th International Conference on Applications of Magnetic Resonance in Food Science held in Copenhagen on the 13–15th September 2004.

Special Publication No. 299

ISBN 0-85404-648-8

A catalogue record for this book is available from the British Library

Published by The Royal Society of Chemistry,
Thomas Graham House, Science Park, Milton Road,
Cambridge CB4 0WF, UK

Registered Charity Number 207890

For further information see our web site at www.rsc.org

Printed by Athenaeum Press Ltd, Gateshead, Tyne and Wear, UK

Preface

This book consists of 26 papers based on lectures presented at the 7th International Conference on Applications of Magnetic Resonance in Food Science. The three-day conference, which was held for the first time in Denmark at The Royal Veterinary and Agricultural University in Copenhagen, attracted over one hundred participants from 20 different countries. The meeting continued the tradition of a friendly gathering with as much, or more, interchange of information during the social events as during the formal sessions. The conference was divided into 5 symposia covering (i) Sensory science, aroma and flavour, (ii) Authenticity and quantification of food, (iii) Functionality, structure and ingredients, (iv) Applications of solid-state NMR methods and (v) New NMR methods and instrumentation. The book follows the form of the conference, although some of the chapters do not necessarily suit the categories.

This year's meeting was subtitled *The Multivariate Challenge* in order to focus on the way in which the interpretation and quantification of magnetic resonance data of complex food systems increasingly requires the application of multivariate data analytical protocols. Since the discovery of the phenomenon of MR spectroscopy in the mid-1940's this technique has developed into one of the most versatile and information-rich analytical techniques. The possible applications range from probing free radicals to studies of dynamic processes in solution, from routine structure elucidation of small molecules to conformational analysis of macromolecules and from quantitative screening of authenticity to MR images of the human brain being impressed by flavours etc. Inspired by the rapidly expanding research field of *metabonomics*, which combines high throughput MR systems with *chemometrics*, it would appear as if the MR food application area is undergoing a transformation towards screening of bioactive components and exploration of functional factors in food. New high throughput instruments are now working intensively on measuring detailed fingerprints of raw materials, food components and final food systems. On-line systems are even being devised for rapid, non-invasive and total quality control. It would appear that hardware technology has evolved to an extent that data technology is becoming the limiting factor.

This collection of papers shows that MR in food science is at a highly sophisticated level and gives good indications of the continuing development towards analysis of larger and more complex food systems, the functionality of which can only be optimally extracted by advanced pattern recognition techniques. Before we know it, we may have developed *bromatonomics*.

We would like to express our deepest thanks to all the active participants, to the staff at Food Technology for hosting the conference, to the conference speakers whose work is documented by this book and to the RSC for making this book a reality. Special thanks are given to Associate Professor Frans van den Berg and Gilda Kischinovsky for help in editing the book.

Søren Balling Engelsen (Editor)
Department of Food Science, The Royal Veterinary and Agricultural University, Rolighedsvej 30, DK-1958 Frederiksberg C, Denmark

Peter Belton (Editor)
School of Chemical Sciences and Pharmacy, University of East Anglia, Norwich NR4 7TJ, UK

Hans Jørgen Jakobsen (Editor)
Department of Chemistry, University of Aarhus, Langelandsgade 140, DK-8000 Aarhus C, Denmark

1992: Surrey
 1994: Aveiro
  1996: Nantes
   1998: Norwich
    2000: Aveiro
     2002: Paris
      2004: Copenhagen

Contents

Applications of Solid-State Methods

New NMR Methods and Instrumentation

Sensory Science, Aroma and Flavour

fMRI AND THE SENSORY PERCEPTION OF FOOD

Edmund T Rolls

University of Oxford, Department of Experimental Psychology, South Parks Road, Oxford OX1 3UD, England, www.cns.ox.ac.uk, email: Edmund.Rolls@psy.ox.ac.uk
Tel. +44-1865-271348, Fax. +44-1865-310447

1   INTRODUCTION

Study of the brain mechanisms involved in taste and smell is revealing some of the brain processing relevant to understanding how the primate (including human) brain interprets the different tastes, odours and oral textures produced by food (*1*). This paper reviews mainly studies of the human brain using functional magnetic resonance imaging to analyze brain regions activated by different properties of food, and builds on more detailed studies using single neuron recordings in macaques.

In the primary taste cortex in the insular opercular region, and in the secondary taste cortex in the orbitofrontal cortex, there are taste neurons that respond to sweet, salt, bitter, sour, and also umami taste as exemplified by monosodium glutamate and inosine 5'monophosphate (*2-4*). In addition, other neurons respond to water, and others to astringent as exemplified by tannic acid (*5*) and to capsaicin (*6, 7*). A difference between these two cortical areas is that in the secondary taste cortex the response of the neurons to sweet taste (glucose) only occurs if hunger is present, so that the orbitofrontal cortex represents the reward value of the taste. Indeed these neurons implement sensory-specific satiety, the process by which the pleasantness of a food eaten in a meal decreases, but the pleasantness of other foods may remain high (*8-13*). It has been shown that approximately 40% of neurons in the orbitofrontal cortex taste and olfactory areas provide a representation of odour that depends on the taste with which the odour has been associated previously, and that this representation is produced by a slowly acting learning mechanism (*14, 15*). The representation of odour thus moves beyond the domain of physico-chemical properties of the odours to a domain where the ingestion-related significance of the odour determines the representation provided by some neurons. Other neurons in the orbitofrontal cortex respond to the sight of food (*9*), and to sensory information about the texture of food in the mouth including the mouth feel of fat, and for other neurons to the viscosity of what is in the mouth (*6, 7, 16, 17*). The temperature and texture of what is in the mouth is represented in the primary taste cortex together with information about taste (*18*).

Another processing principle in the orbitofrontal cortex is that in addition to unimodal representations of the taste, olfactory, somatosensory, and visual properties of sensory stimuli, some neurons combine inputs from these different modalities, and these combination-selective neurons provides an information-rich representation of a wide range of the sensory qualities of food. One key to understanding these combinatorial representations is learning in such a way that individual neurons come to respond to particular combinations of taste, olfactory, texture, and associated visual stimuli.

A diagram of the taste and related olfactory, somatosensory, and visual pathways in primates is shown in Fig. 1.

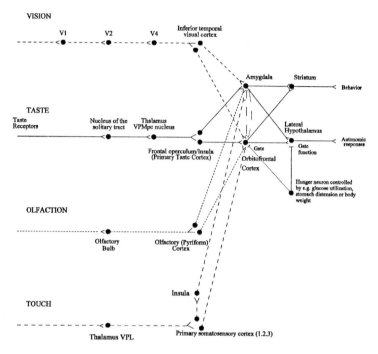

Figure 1   *Schematic diagram of the taste and olfactory pathways in primates showing how they converge with each other and with visual pathways. The gate functions shown refer to the finding that the responses of taste neurons in the orbitofrontal cortex and the lateral hypothalamus are modulated by hunger. VPMpc - ventralposteromedial thalamic nucleus; V1, V2, V4 - visual cortical areas.*

2   IMAGING STUDIES IN HUMANS

2.1   Taste

In humans it has been shown in neuroimaging studies using functional Magnetic Resonance Imaging (fMRI) that taste activates an area of the anterior insula/frontal operculum, which is probably the primary taste cortex, and part of the orbitofrontal cortex, which is probably the secondary taste cortex (*19-21*). The orbitofrontal cortex taste area is distinct from areas activated by odours and by pleasant touch (*19*). It has been shown that within individual subjects separate areas of the orbitofrontal cortex are activated by sweet (pleasant) and by salt (unpleasant tastes) (*20*). Francis et al. (*19*) also found activation of the human amygdala by the taste of glucose. Extending this study, O'Doherty et al.(*20*) showed that the human amygdala was as much activated by the affectively pleasant taste of glucose as by the affectively negative taste of NaCl, and thus provided evidence that the human amygdala is not especially involved in processing aversive as compared to rewarding stimuli. Another study has recently shown that umami taste stimuli, of which an exemplar is monosodium glutamate (MSG) and which capture what is described as the taste of protein, activate similar cortical regions of the human taste system to those activated by a prototypical taste stimulus, glucose (see Fig. 2) (*22*). A part of the rostral anterior cingulate cortex (ACC) was also activated. When the nucleotide 0.005 M inosine 5'-monophosphate (IMP) was added to MSG (0.05 M), the BOLD (blood oxygenation-level dependent) signal in an anterior part of the orbitofrontal cortex showed supralinear additivity, and this may reflect the subjective enhancement of umami taste that has been described when IMP is added to MSG. Overall, these results illustrate that the responses of the brain can reflect inputs produced by particular combinations of sensory stimuli with supralinear activations, and that the combination of sensory stimuli may be especially represented in particular brain regions.

2.2 Odour

In humans, in addition to activation of the pyriform (olfactory) cortex (*23-25*), there is strong and consistent activation of the orbitofrontal cortex by olfactory stimuli (*19, 26*). In an investigation of where the pleasantness of olfactory stimuli might be represented in humans, O'Doherty et al (*27*) showed that the activation of an area of the orbitofrontal cortex to banana odour was decreased (relative to a control vanilla odour) after bananas were eaten to satiety. Thus activity in a part of the human orbitofrontal cortex olfactory area is related to sensory-specific satiety, and this is one brain region where the pleasantness of odour is represented.

We have also measured brain activation by whole foods before and after the food is eaten to satiety (*22*). The aim is to show using a food that has olfactory, taste and texture components the extent of the region that shows decreases when the food becomes less pleasant, in order to identify the different brain areas where the pleasantness of the odour, taste and texture of food are represented. The foods eaten to satiety were either chocolate milk, or tomato juice. A decrease in activation by the food eaten to satiety relative to the other food was found in the orbitofrontal cortex (*28*) but not in the primary taste cortex. This study provided evidence that the pleasantness of the flavour of food is represented in the orbitofrontal cortex.

An important issue is whether there are separate regions of the brain discriminable with fMRI that represent pleasant and unpleasant odours. To investigate this, we measured the brain activations produced by three pleasant and three unpleasant odours. The pleasant odours chosen were linalyl acetate (floral, sweet), geranyl acetate (floral) and alpha-ionone (woody, slightly food-related). (Chiral substances were used as racemates.) The unpleasant odours chosen were hexanoic acid, octanol and isovaleric acid. We found that they activated dissociable parts of the human brain (*29*). Pleasant but not unpleasant odours were found to activate a medial region of the rostral orbitofrontal cortex. Further, there was a correlation between the subjective pleasantness ratings of the six odours given during the investigation with activation of a medial region of the rostral orbitofrontal cortex. In contrast, a correlation between the subjective unpleasantness ratings of the six odours was found in regions of the left and more lateral orbitofrontal cortex. Activation was also found

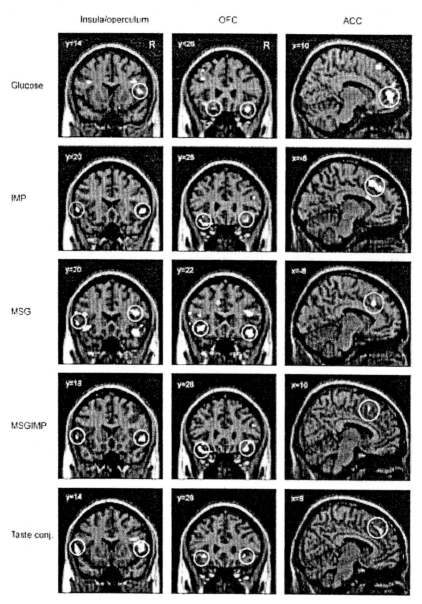

Figure 2   *Activation of the human primary taste cortex in the insula/frontal operculum; the orbitofrontal cortex (OFC); and the anterior cingulate cortex (ACC) by taste. The stimuli used included glucose, two umami taste stimuli (monosodium glutamate (MSG) and inosine monophosphate (IMP)), and a mixture of the two umami stimuli. Taste conj refers to a conjunction analysis over all the taste stimuli. (After 22).*

in the anterior cingulate cortex, with a middle part of the anterior cingulate activated by both pleasant and unpleasant odours, and a more anterior part of the anterior cingulate cortex showing a correlation with the subjective pleasantness ratings of the odours.

These results provide evidence that there is a hedonic map of the sense of smell in brain regions such as the orbitofrontal cortex and cingulate cortex. It will be of interest in future studies to extend the range of odors to include perfumes and flavours, to determine how they map onto the areas just described. It will also be of interest to investigate where flavours formed by a combination of olfactory and taste input are represented, and how cognitive factors affect the brain activations produced by odours.

The topological representation of the hedonic properties of sensory stimuli such as smell in the orbitofrontal cortex can be understood with some of the fundamental principles of computational neuroscience (*30, 31*) as follows. Given the evidence described above that the reward-related or affective properties of sensory stimuli, rather than for example the intensity of the stimuli, is represented in the orbitofrontal cortex, a topological map of the hedonic value of stimuli is produced in which neurons that have similar hedonic value are placed close together in the map. This self-organizing map results from processes that occur in competitive networks, the building blocks of sensory systems, in which the neurons are coupled by short range (~ 1 mm) excitation (implemented by the recurrent excitatory connections between cortical pyramidal cells) and longer range inhibition (implemented by inhibitory interneurons) (see *30, 31*). Such self-organizing maps are a useful feature on brain connectivity, for they help to minimize the length of the connections between neurons that need to exchange information to perform their computations. It is for this reason, it is hypothesized, that neurons with similar hedonic value are placed close together. In the case of olfactory stimuli, this results in an activation region for pleasant olfactory stimuli (in the medial orbitofrontal cortex) and a separate activation region for unpleasant olfactory stimuli (more laterally in the orbitofrontal cortex). Of course, part of the support for such a map (i.e. a factor which helps different types of stimuli to be separated in the map) may arise because unpleasant olfactory stimuli may be more generally associated with other sensory inputs such as trigeminal inputs. Understanding this principle may be very valuable in helping to interpret the results of neuroimaging experiments.

2.3   Olfactory-taste convergence to represent flavour

To investigate where in the human brain interactions between taste and odour stimuli may be realised to implement flavour, we performed an event-related fMRI study with sucrose and MSG taste, and strawberry and methional (chicken) odours, delivered unimodally or in different combinations (*32*). The brain regions that were activated by both taste and smell included parts of the caudal orbitofrontal cortex, amygdala, insular cortex and adjoining areas, and anterior cingulate cortex. It was shown that a small part of the anterior (putatively agranular) insula responds to unimodal taste and to unimodal olfactory stimuli; and that a part of the anterior frontal operculum is a unimodal taste area (putatively primary taste cortex) not activated by olfactory stimuli. Activations to combined olfactory and taste stimuli where there was little or no activation to either alone (providing positive evidence for interactions between the olfactory and taste inputs) were found in a lateral anterior part of the orbitofrontal cortex. Correlations with consonance ratings for the smell and taste combinations, and for their pleasantness, were found in a medial anterior part of the orbitofrontal cortex. These results provide evidence on the neural substrate for the convergence of taste and olfactory stimuli to produce flavour in humans, and where the pleasantness of flavour is represented in the human brain.

2.4   Oral viscosity and fat texture

The viscosity of food in the mouth is represented in the human primary taste cortex (in the anterior insula), and also in a mid-insular area that is not taste cortex, but which represents oral somatosensory stimuli (*33*). In these regions, the fMRI BOLD activations are proportional to the log of the viscosity of carboxymethyl cellulose in the mouth. Oral viscosity is also represented in the human orbitofrontal and perigenual cingulate cortices, and it is notable that the perigenual cingulate cortex, an area in which many pleasant stimuli are represented, is strongly activated by the texture of fat in the mouth and also by oral sucrose (*33*).

3 EMOTION

The brain areas where the pleasantness or affective value of smell and taste are represented are closely related to the brain areas involved in emotion. Emotions can usefully be defined as states elicited by rewards and punishers (*1a*), and olfactory and taste stimuli can be seen as some of the classes of stimuli that can produce emotional states. Part of the importance of the orbitofrontal cortex in emotion is that it represents some primary (or unlearned) rewards and punishers, such as taste and pleasant touch (*19, 34a*), and also learns the association between previously neutral stimuli and primary reinforcers. This type of learning is called stimulus-reinforcement association learning, and is the type of learning that is fundamental in learned emotional states. In addition to reinforcers such as taste, odour, and touch, quite abstract emotion-producing stimuli are represented in other parts of the orbitofrontal cortex. For example, the medial orbitofrontal cortex is activated in humans according to how much money is won in a probabilistic reward / punishment task, and the lateral orbitofrontal cortex is activated according to how much money is lost in the same task (*35*).

It is thus becoming possible to start to understand not only where the affective value of smell and taste is represented in the brain, but also how these representations fit into a wider picture of the brain processes underlying emotion.

4 CONCLUSION

The primate orbitofrontal cortex is an important site for the convergence of representations of the taste, smell, sight and mouth feel of food, and this convergence allows the sensory properties of each food to be represented and defined in detail. The primate (including human) orbitofrontal cortex is also the region where a short-term, sensory-specific, control of appetite and eating is implemented. Moreover, it is likely that visceral and other satiety-related signals reach the orbitofrontal cortex and there modulate the representation of food, resulting in an output that reflects the reward (or appetitive) value of each food. The orbitofrontal cortex contains not only representations of taste and olfactory stimuli, but also of other types of rewarding and punishing stimuli including the texture and temperature of food, and pleasant touch, and all these inputs, together with the functions of the orbitofrontal cortex in stimulus-reward and stimulus-punishment association learning, provide a basis for understanding its functions in emotional and motivational behaviour (see *1, 36-39*). Moreover, the orbitofrontal cortex shows responses that reflect combinations of sensory inputs, and help us to understand the ways in which sensory inputs are combined, sometimes non-linearly, to produce complex representations reflecting combinations of particular sensory inputs.

References

1. E. T. Rolls, *The Brain and Emotion* (Oxford University Press, Oxford, 1999).
2. E. T. Rolls, S. Yaxley, Z. J. Sienkiewicz, *Journal of Neurophysiology* **64**, 1055-1066 (1990).
3. E. T. Rolls, *Critical Reviews in Neurobiology* **11**, 263-287 (1997).
4. E. T. Rolls, T. R. Scott, in *Handbook of Olfaction and Gustation* R. L. Doty, Ed. (Dekker, New York, 2003), vol. chap, 32, pp. 679-705.
5. H. D. Critchley, E. T. Rolls, *Chemical Senses* **21**, 135-145 (1996).
6. E. T. Rolls, J. V. Verhagen, M. Kadohisa, *Journal of Neurophysiology* **90**, 3711-3724 (2003).
7. M. Kadohisa, E. T. Rolls, J. V. Verhagen, *Neuroscience* **127**, 207-221 (2004).
8. E. T. Rolls, Z. J. Sienkiewicz, S. Yaxley, *European Journal of Neuroscience* **1**, 53-60 (1989).
9. H. D. Critchley, E. T. Rolls, *Journal of Neurophysiology* **75**, 1673-1686 (1996).
10. B. J. Rolls, E. T. Rolls, E. A. Rowe, K. Sweeney, *Physiology and Behavior* **27**, 137-142 (1981).
11. B. J. Rolls *et al.*, *Physiology and Behavior* **26**, 215-221 (1981).
12. B. J. Rolls, P. M. Van Duijvenvoorde, E. T. Rolls, *Appetite* **5**, 337-348 (1984).
13. E. T. Rolls, J. H. Rolls, *Physiology and Behavior* **61**, 461-473 (1997).
14. E. T. Rolls, L. L. Baylis, *Journal of Neuroscience* **14**, 5437-5452 (1994).
15. E. T. Rolls, H. Critchley, E. A. Wakeman, R. Mason, *Physiology and Behavior* **59**, 991-1000 (1996).
16. E. T. Rolls, H. D. Critchley, A. S. Browning, A. Hernadi, L. Lenard, *Journal of Neuroscience* **19**, 1532-1540 (1999).
17. J. V. Verhagen, E. T. Rolls, M. Kadohisa, *Journal of Neurophysiology* **90**, 1514-1525 (2003).
18. J. V. Verhagen, M. Kadohisa, E. T. Rolls, *Journal of Neurophysiology* **in press** (2004).
19. S. Francis *et al.*, *Neuroreport* **10**, 453-459 (1999).
20. J. O'Doherty, E. T. Rolls, S. Francis, R. Bowtell, F. McGlone, *Journal of Neurophysiology* **85**, 1315-1321 (2001).
21. I. E. T. De Araujo, M. L. Kringelbach, E. T. Rolls, F. McGlone, *Journal of Neurophysiology* **90**, 1865-1876 (2003).
22. I. E. T. De Araujo, M. L. Kringelbach, E. T. Rolls, P. Hobden, *Journal of Neurophysiology* **90**, 313-319 (2003).
23. A. Poellinger *et al.*, *Neuroimage* **13**, 547-560 (2001).
24. N. Sobel *et al.*, *Journal of Neurophysiology* **83**, 537-551 (2000).
25. D. H. Zald, J. V. Pardo, *Proceedings of the National Academy of Sciences USA* **94**, 4119-4124. (1997).
26. R. J. Zatorre, M. Jones-Gotman, A. C. Evans, E. Meyer, *Nature* **360**, 339-340 (1992).
27. J. O'Doherty *et al.*, *Neuroreport* **11**, 893-897 (2000).
28. M. L. Kringelbach, J. O'Doherty, E. T. Rolls, C. Andrews, *Cerebral Cortex* **13**, 1064-1071 (2003).

29. E. T. Rolls, M. L. Kringelbach, I. E. T. de Araujo, *European Journal of Neuroscience* **18**, 695-703 (2003).
30. E. T. Rolls, A. Treves, *Neural Networks and Brain Function* (Oxford University Press, Oxford, 1998).
31. E. T. Rolls, G. Deco, *Computational Neuroscience of Vision* (Oxford University Press, Oxford, 2002).
32. I. E. T. De Araujo, M. L. Kringelbach, E. T. Rolls, *European Journal of Neuroscience* **18**, 2374-2390 (2003).
33. I. E. T. De Araujo, E. T. Rolls, *Journal of Neuroscience* **24**, 3086-3093 (2004).
34. E. T. Rolls *et al.*, *Cerebral Cortex* **13**, 308-317 (2003).
35. J. O'Doherty, M. L. Kringelbach, E. T. Rolls, J. Hornak, C. Andrews, *Nature Neuroscience* **4**, 95-102 (2001).
36. E. T. Rolls, *Journal of Nutrition* **130**, S960-S965 (2000).
37. E. T. Rolls, *Cerebral Cortex* **10**, 284-294 (2000).
38. E. T. Rolls, in *Neural and Metabolic Control of Macronutrient Intake* H.-R. Berthoud, R. J. Seeley, Eds. (CRC Press, Boca-Raton, Florida, 2000) pp. Chap 17, 247-262.
39. E. T. Rolls, in *The Amygdala: A Functional Analysis* J. P. Aggleton, Ed. (Oxford University Press, Oxford, 2000) pp. 447-478.

Acknowledgements

The author has worked on some of the experiments described here with Drs. R. Bowtell, A. Browning, H. Critchley, I.E.T. de Araujo, S. Francis, M.L. Kringelbach, M. Kadohisa, G. Kobal, F. McGlone, J. O'Doherty, T. R. Scott, Z. J. Sienkiewicz, E. A. Wakeman, J. Verhagen, L. L. Wiggins (L.L.Baylis) and S. Yaxley and their collaboration is sincerely acknowledged. Some of the research from the author's laboratory was supported by the Medical Research Council.

HR NMR TO STUDY QUALITY CHANGES IN MARINE BY-PRODUCTS

E. Falch[1,2], T. R. Størseth[1] and M. Aursand[1]

[1] SINTEF Fisheries and Aquaculture, Trondheim, Norway
[2] The Norwegian University of Science and Technology, Department of Biotechnology, Trondheim, Norway

1 INTRODUCTION

Marine raw materials contain health beneficial lipids[1,2] with applications in food, healthcare and pharmaceutical products. However, marine lipids are highly susceptible to lipid oxidation and to chemical reactions caused by endogenous enzymes. The lipid composition in fish tissues is a complex mixture of fatty acids esterified in neutral and polar lipids, but also sterols, vitamins and other lipid components are found. Lipids in fish muscle are relatively well characterised. However, more effort is needed to characterise the chemical composition of potential by-products such as head, trimmings and visceral fractions to make utilisation of the total catch of fish feasible[3]. Recently, a number of studies on lipid composition in cod by-products are reported [4,5,6,7,8,9].

Several analytical methods to determine lipid composition are available. Traditional methods are generally based on lipid extraction which may be prone to incomplete extraction of the desired components. High resolution (HR) NMR may provide chemical information about all NMR visible lipid classes and thereby changes in lipids and their reaction products. HR- NMR has been used to characterise a wide range of chemical compounds found in marine lipids[10]. The omega-3 content in muscle lipids extracted from Atlantic salmon[11] and in different fish oils[12] as well as the content of docosahexaenoic acid in different fish oils have been quantified using ^{1}H NMR. Moreover, ^{1}H-NMR has also been successfully used to study lipid oxidation in ethyl docosahexaenoate[13]. The ^{13}C spectra may provide detailed information about the fatty acid composition, positional distribution of fatty acids in acylglycerolmolecules[14,15,16] and phospholipids[15]. The phospholipids can be identified by ^{1}H and ^{13}C NMR due to their differences in the structure of the polar head groups, whereas ^{31}P NMR has been more widely used to differentiate individual phospholipids including lysophospholipids[17].

Magic angle spinning (MAS) NMR was first applied on lipids in 1972[18]. Since then, the method has been gradually improved. MAS experiments can be performed on intact tissues and spinning the samples at the angle of 54.7°. Thus, partial averaging of chemical shift anisotropies and dipolar couplings thereby significantly reduce the line broadening effect[19]. The method has been used successfully to study different biological tissues and intact rat liver. The obtained spectra were well resolved providing detailed information about a wide range of chemical compounds in these materials[20].

2   MATERIALS AND METHODS

2.1 Raw material

Among the by-products from cod (*Gadus morhua*), roe, milt and liver are presently commercial food products. Roe and milt are generally rich in phospholipids. In our study, the lipids in milt and roe contained 60-77% phospholipids, 17-18% cholesterol, and 0-9% triacylglycerols. Examples of endogenous seafood enzymes catalysing hydrolysis of phospholipids are shown in Fig. 1 whereas the formation of lysophosphatidylcholine and free fatty acids from phosphatidylcholine is illustrated in Fig. 2.

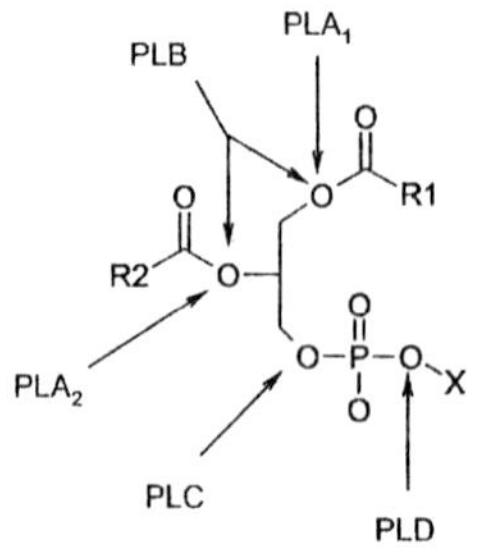

Figure 1   *Various enzymes catalysing phospholipid hydrolysis. PLA₁, PLA₂ and PLB are phospholipases belonging to the group acyl hydrolases together with lysophospholipases while PLC and PLD are phosphodiesterases. R₁ and R₂ denotes fatty acids and X represents the polar head group[21]*

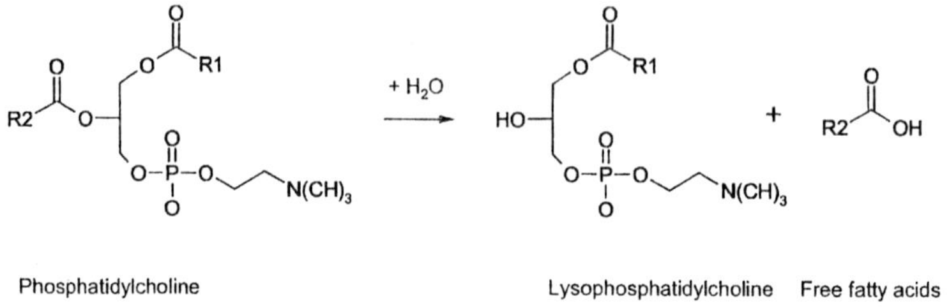

Phosphatidylcholine   Lysophosphatidylcholine   Free fatty acids

Figure 2   *Hydrolysis of phosphatidylcholine[22].*

The lipids in cod liver consist primarily of triacylglycerols (>95% in our study)[4] with a level of eicosapentaenoic acid (EPA) and docosahexaenoic acid (DHA) at respectively 7% and 11% respectively. Comparatively, lipids in roe and milt contain 15% EPA and 25% DHA.

The by-products were collected from a batch of fresh cod and immediately frozen and stored at -80°C until NMR analysis. Immediately after gutting a storage experiment was started with fresh milt packed in dark plastic bags (10g) and stored at 4°C up to one week. Samples were withdrawn at time specific intervals. Lipids from samples at time 0 and after one week (extracted by the method of Bligh and Dyer[23]) were analysed by thin layer chromatography[24]. The lipid extracts (100 mg) were also transferred into 5 mm NMR tubes and dissolved in 50 ml deuterated chloroform ($CDCl_3$) for ^1H, ^{13}C and 2D ($^1H,^1H$ COSY) NMR analysis.

Additionally, samples at time 0 and after 1 week storage were analysed by HR-MAS at 4°C. The samples were weighed in the rotor (50μl) and added deuterium oxide (D_2O) with trimethylsilylpropionic acid (TSP) as a reference.

2.2 NMR procedures

The NMR experiments were carried out on a Bruker DRX 600 spectrometer (Bruker Biospin GmBH, Germany) resonating at 600 MHz (proton frequency).

For liquid state experiments, a 5 mm BBO probe was used at 298 K. For HR-MAS experiments, a 4 mm MAS probe was used with a MAS spin rate of 5 kHz. All ^1H experiments were recorded using a 90° pulse, 256 scans with 64 K data points were recorded over a spectral width of 7 kHz.

In the HR-MAS spectra the water resonance was suppressed by a pulse sequence using presaturation in the interscan delay. Zero filling and 0.3 Hz exponential line broadening was applied before Fourier transformations. ^{13}C spectra were recorded using a 90° pulse. 8 k scans were accumulated with 64 k data points with a spectral width of 30303 Hz and 3Hz exponential line broadening was applied before Fourier transformation. The $^1H,^1H$-COSY NMR measurements were carried out using a 45° polarization transfer pulse to record 2048/1024 data points for the F2/F1 directions. Before Fourier transformation both dimensions were zero-filled and apodized by a squared sine bell function. The HR-MAS $^1H,^1H$-COSY spectra were recorded using presaturation of the water resonance in the interscan delay. Chemical shifts were referenced to TSP in the HR-MAS experiments where the samples were added D_2O as a field frequency lock. A series of 1D (^1H and ^{13}C) and 2D (^1H-^1H COSY) spectra were acquired at 4°C by HR-MAS over a 3 day period.

3   RESULTS AND DISCUSSION

The levels of phospholipids were halved during one week of storage at 4°C. The levels of cholesterol were initially high but decreased during storage. Simultaneously, the levels of steryl esters and free fatty acids increased from 0 up to significant levels during storage (Fig. 3).

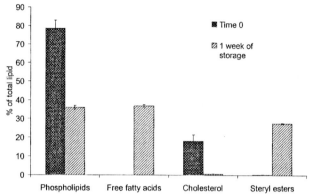

Figure 3   *Lipid classes in milt samples (% of total lipid content).*

^{13}C NMR spectra of lipid extracts provided particularly valuable information about the changes occurring. By studying the whole ^{13}C NMR spectra of milt extracts at time 0 and after 1 week storage, the differences in the amount of cholesterol were clearly recognizable (Fig. 4).

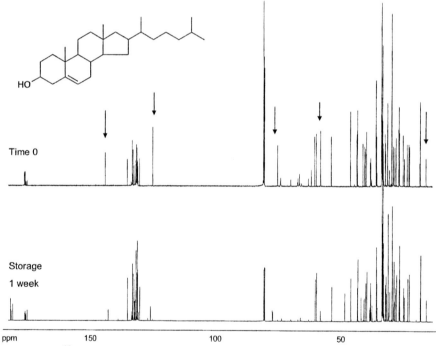

Figure 4 ^{13}*C NMR spectra of milt lipids extracts of milt at time 0 and after 1 week storage at 4°C. The structure of cholesterol which is a major constituent in this material and some peaks from the molecule are illustrated.*

The signals from cholesterol carbons were detected as narrow singlets. It was obvious that the hydroxyl group of cholesterol had changed since the signal of the C3 carbon (71.81 ppm) decreased. Also, the disappearance of double bound signals from cholesterol (121.7 and 140.8 ppm (singlets)) were found while new peaks appeared in the olefinic region. Changes due to esterification of cholesterol were observed in the ^1H spectra around 1.0 ppm (Fig.5) and at 3.8 and 4.5 ppm (Fig. 8).

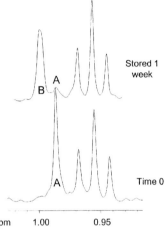

Figure 5   *^1H NMR spectra of milt lipids extracted at time 0 and after 1 week storage at 4°C. The ratio between peak A and B as affected by storage and are similar to reported cholesterol and cholesteryl esters chemical shift values[25].*

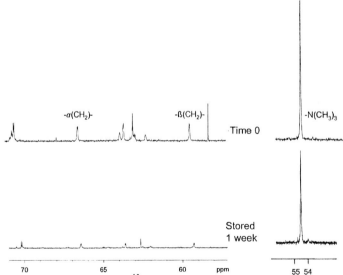

Figure 6   *Glycerol region of the ^{13}C NMR spectra of milt lipids at time 0 and after 1 week storage at 4°C. The phospholipids region is shown. The assigned peaks originate from carbons in the phosphogroup of the phospholipids.*

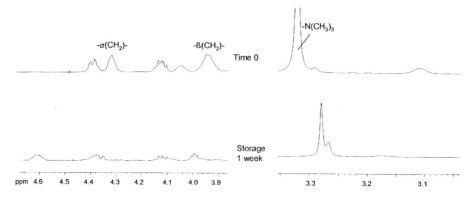

Figure 7 ^1H *NMR spectra of milt lipids extracted at time 0 and after 1 week storage at 4°C. The signals originate from hydrogens in the phosphogroups of phospholipids.*

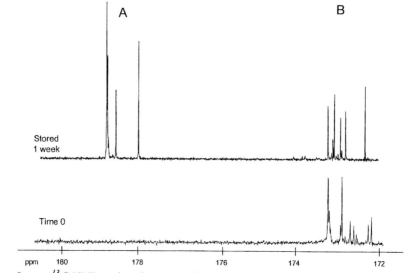

Figure 8 ^{13}C *NMR carbonyl spectra of milt lipid extracted at time 0 and after 1 week storage at 4°C. (A)Free fatty acid region, and (B) region containing information about positional distribution of unsaturated fatty acids in the glycerol molecule.*

The free fatty acids in the lipid extracts of milt were significant higher in the samples stored for one week compared to the samples at time 0 which were at non detectable levels. The spectra were also well resolved concerning the carbonyl region. This differentiated between the fatty acids released based on their levels of unsaturation. ^{13}C NMR spectra also provided information about the glycerol position (in the tri, di monoacyl phospholipids or in the lysophospholipid) from which the fatty acids were released. Our spectra showed that in the milt samples, the EPA and DHA are mainly located at glycerol-carbon atom 2 (sn-2) position and that a major fraction of these fatty acids were hydrolysed during storage. By studying the glycerol region (Fig. 6), a decrease of signals from phospholipid carbons was found without a corresponding increase due to lysophospholipids signals at similar levels. This may indicate that also other enzymes than phospholipases were active hydrolysing the polar head groups of phospholipids. The ^{13}C NMR spectra provided information about the specific fatty acids, position of fatty acids in the triacylglycerol and phospholipids molecules. Fig. 8 illustrates the formation of free fatty acids during 1 week storage. The ^{1}H NMR spectra provided information about changes caused by hydrolysis of phospholipids (3.2 – 4.6 ppm) (Fig. 7).

HR-MAS analysis provided well resolved ^{1}H NMR spectra (Fig. 9 presents results from liver). In milt, time dependent differences in region at 3.0-4.5 ppm were found after relative short time. Possible influence of high spinning rate, pH variations and peak assignments when using MAS should be investigated further before drawing any further conclusions.

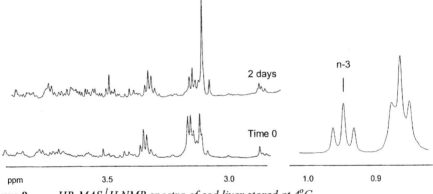

Figure 9 *HR-MAS ^{1}H NMR spectra of cod liver stored at 4°C.*

The marine lipids in by-products were labile and different biochemical processes occur simultaneously during chilled storage. It has previously been reported that due to difficulties in isolation and characterisation of lipases in fish tissues, limited information about the enzymology of digestive lipases are available[26]. NMR can be a valuable tool for such studies. The level of free fatty acids – due to lipid hydrolysis - is widely used industrially as an indicator of lipid quality. However, our results show this might not be the whole answer. Clearly, there was an increase of the levels of free fatty acids in milt during one week of storage. The strong indication of esterification of free fatty acids to the cholesterol molecule might be a hidden effect since the degree of lipolysis was higher than expected. NMR could also provide detailed molecular information about new compounds being formed. These compounds may play a significant role regarding potential health effects and sensory properties of food.

Reaction products from the lipid oxidation and lipolysis (free fatty acids) are known to affect the sensory acceptance of the products. Possible health effects due to the reaction products being formed should also be given attention. Some of oxidation compounds are toxic at high concentrations and studies have shown that hydroperoxides and aldehydes might cause damage of DNA[27].

4 CONCLUSIONS

Results from the early stages of this work have shown that a combination of various NMR techniques may provide a more detailed understanding of the chemical and biochemical reactions occurring in the lipids of marine by-products. Detailed information about lipolysis and other changes in lipid classes were found. However, the used methods need further optimisation including more extensive assignments of unknown peaks.

Acknowledgements

The project was funded by Norwegian Research Council (project: *Increased value adding from by-products and by-catches) and EU (QLK1-CT2000-01017) Utilisation and stabilization of by-products from cod species.*

References:

1   Dyerberg, J., Bang H. O., Stofferson, E., Monkada S., and Vane, J.R., *Lancet*, 1978, **2**, 117.

2   Vanschoonbeek, K., de Maat, M.P., Heemskerk, J.W., *J. Nutr. 2003*, **133**, 657.

3   Rustad, T, and E. Falch, *Food Science & Technology*, The International Quarterly of the Institute of Food Science and Technology, 2002, **6** (2), p 36-37, 39.

4   Falch, E., M. Aursand, and T. Rustad, By-products from cod species (*Gadidae*) as a source of marine lipids. 2004, Submitted manuscript.

5   Falch, E., R. Jónsdóttir, S. Arason, K. A. Þórarinsdóttir, N. B. Shaw, J. P. Kerry, C. Malone, J.P. Berge, J. Dumay, J. Rainuzzo, M. Sandbakk, M. Aursand, Oral presentation and proceedings at the Trans-Atlantic Fisheries Technology conference, 2003, Iceland.

6   Stoknes, I. S., H.M.W Økland, E. Falch, and M. Synnes, *Comparative Biochemistry and Physiology*, Part B, 2004, **38**, 183.

7   Slizyte, R, E. Dauksas, E. Falch, T. Rustad and I. Storrø, *Process Biochemistry*, 2004, In press.

8   Slizyte, R, E. Dauksas, E. Falch, I. Storrø and T. Rustad, *Process Biochemistry*, 2004, Accepted

9   E. Dauksas, E. Falch,, R. Slizyte, and T. Rustad, *Process Biochemistry*, 2004, Submitted

10 Aursand, M., R. Rainuzzo, H. Grasdalen, *Comp. Biochem. Physiol.*, 1995, **112B**, 315.

11 Aursand, M., R. Rainuzzo, H. Grasdalen, *J. Am. Oil Chem. Soc.* 1993, **70**(10), 971.

12 Igarashi T., M. Aursand, Y. Hirata, I.S. Gribbestad, S. Wada, and M. Nonaka, *J. Am. Oil Chem. Soc.* 2000, **77**(7), 737.

13 Falch, E., H. Anthonsen, D. Axelson, and M, Aursand, *JAOCS*, 2004, *J. Am. Oil Chem. Soc.* Accepted for publication

14 Mannina L., C. Luchinat, M. C. Emanuelle, and A. Segre, *Chem. Phys. Lipids*, 1999, **103** (1-2), 47.

15 Medina, I., R. Sacchi, I. Giudicianni, and S. Aubourg, *J. Am. Oil. Chem. Soc*, 1998, **75** (2),147.

16 Aursand, M. Jørgensen, L., and Grasdalen, H., *J. Am. Oil Chem. Soc.*, 1995, **72**, 293.

17 Schiller, J. and K. Arnold, *Med. Sci. Monit.* 2002, **8**, MT205-MT222.

18 Everts, S. and J. H. Davies, *Biophysical Journal*, 2000, **79**, 885-897

19 Andrew, E.R. and R. G. Eades, *Nature*, **183**, 1802.

20 Bollard, M. E., S. Garrod, E. Holmes, J. C., Lindon, E. Humpfer, M. Spraul, and J. K. Nicholson, *Magnetic Resonance in Medicine,* 2000, **44**, 201.

21 Waite, M. The phospholipases, Handbook of Lipid Reaserch, 5, Plenum Press, 1987, NY. 332 p.

22 Hanahan, D.J. A Guide to Phospholipid Chemistry, Oxford University Press, 1997, NY. 214p.

23 Bligh, E. G. and W. J. Dyer, *Can. J. Biochem. Physiol,* 1959, **37** (8) 911.

24 Aursand, M., and Grasdalen, H. *Chem. Phys. Lipids*, 1992, **62**, 239.

25 Tosi, M. R., G. Bottura, P. Lucchi, A. Reggiani, A. Trinchero, and V. Tugnoli, *International Journal of Molecular Medicine, 2003,* **11**, 92.

26 Ono, H., N. Iijima, *Fish Physiol Biochem*, 1998, **18**, 135.

27 Yang, M-H. and K. Schaich, *Free Radical Biology & Medicine*, 1996, **20**(2), 225.

ON THE USE OF LOW-FIELD NMR METHODS FOR THE DETERMINATION OF
TOTAL LIPID CONTENT IN MARINE PRODUCTS

Geir Humborstad Sørland[1], Per Magnus Larsen[1], Frank Lundby[2], Henrik W. Anthonsen[3],
Bente Jeanette Foss[3]

[1]Anvendt Teknologi A/S, Hagebyv. 32, N-9404 Harstad, Norway, [2]MATFORSK,
Osloveien 1, N-1430 Ås, Norway, [3]Dept. of Chemistry, Norwegian University of Science
and Technology, N-7491 Trondheim, Norway

1 INTRODUCTION

NMR applications for the determination of lipid content are currently available as
laboratory methods [1-3]. The applications are run on permanent magnets at an operating
magnetic field of approximately 0.5 Tesla. At this magnetic field proton chemical shift
information is not available, and the NMR signal from the lipids must be resolved from the
other components by other means. This could be differences in transverse (T_2) or
longitudinal (T_1) relaxation times or molecular mobility (diffusion).

A widely used method for dried samples is based on the lipid being the only
component with a relatively long transverse relaxation time [2]. The recording of one spin
echo signal at an observation time of approximately 0.3 ms is then assigned to the fat
signal from the sample. Any possible lipid component with short T_2 would then not be
completely accounted for. Another method applies pulsed field gradients in a multi pulsed
field gradient spin echo experiment (combined PFG-CPMG), where the moisture signal is
suppressed to an insignificant amount at the first echo while the attenuation of the lipid
signal due to diffusion and relaxation effects is recorded [1, 4]. The attenuation of the lipid
signal is then fitted to an exponential decay resulting in a fitted initial lipid signal where
loss of signal due to diffusion and relaxation effects has been corrected for.

While both methods seem to work well on initially wet samples, there is a major
group of systems were both methods fail to determine the total fat content. On powdered
samples where the moisture content is small (~10%) and the phospholipid content is
relatively high, a lipid component with short T_2 is not completely accounted for by either
of the methods. The spin echo method fails, because significant amounts of the signal from
the phospholipids have already decayed at the first measuring point. Thus the measured fat
content will be underestimated. The combined PFG-CPMG method fails because the first
spin echo is recorded at approximately 10 ms to avoid interference from protein signal
from the interface between protein and moisture [5]. At an observation time of 10 ms the
phospholipid from the powdered sample with small moisture content has decayed to an
insignificant amount, and an extrapolation back to zero observation time will not recapture
the initial amount of phospholipids.

2 THEORY

The resolution of a lipid signal becomes feasible when one can isolate the lipid NMR signal from the components as protein, carbohydrates and moisture. Without chemical shift information using low resolution NMR equipment this can be achieved by making use of significant differences in the dynamic behaviour of the components. To resolve the lipid signal from the other components the NMR spin echo technique is used either in combination with pulsed magnetic field gradients or not.

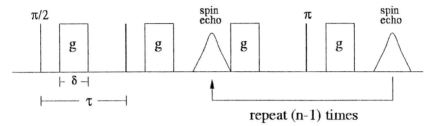

Figure 1 *The combined Pulsed Field Gradient Spin Echo (PFG-CPMG) experiment*

In general the attenuation from a CPMG with the option of applied pulsed magnetic field gradients (Figure 1) within the τ-intervals is written

$$I = \sum_k \sum_j I_{j,k} e^{-\frac{2n\tau}{T_2^{j,k}} - \frac{2n\tau^3}{3}\gamma^2 G_i^2 D_{j,k}} e^{-n\gamma^2 g^2 D_{j,k}\delta^2(\tau-\frac{\delta}{3})} \tag{1}$$

where Σ_k is the sum over the components present in the sample while Σ_j reflects the multi exponential behaviour within each component. $I_{j,k}$ is the initial intensity, $D_{j,k}$ is the diffusion coefficient, $T_2^{j,k}$ is the transverse relaxation time, G_i is the strength of the internal magnetic field gradient, g is the strength of the applied magnetic field gradient, γ is the gyromagnetic ratio, δ is the gradient pulse length, τ is the time interval between 90 degree RF pulse and 180 degree RF pulse, and **n** is the echo number (**n** = 1,2,3,etc.).

2.1 Resolving lipids in powdered systems

In powdered samples there is usually 5-15 % moisture. With an overlap in transversal relaxation times (T_2) between moisture bound to the solid matrix and possible amounts of lipids with short T_2's (bound lipid), it is difficult to isolate the bound lipid signal from the moisture. To isolate the total lipid signal from any other component it is therefore crucial to dry out the ~10 % of moisture prior to the NMR measurements.

When drying the moisture out of the sample the protein and carbohydrates are solids with a T_2 of the order of 10 us. If the first echo in the CPMG is recorded at more than 100 us, the signal from these components will have decayed to an insignificant amount. However, as the bound lipid also exhibits a relatively short T_2, it is important to start the recording of the CPMG train at the shortest observation time possible and record as many points as possible during the first few milliseconds. As will be verified in the experimental section, we have found that it is possible to resolve the lipid components from the other components within a dried sample by using a τ-value of 40 μs and start the recording of the CPMG train at echo number 2 which is at an observation time of 160 μs. With a 180 degree RF pulse duration of 9 μs, the protein and carbohydrate signal does not contribute to the echo signal at 160 μs even though there are $T_{1\rho}$ effects at such short τ-values [6].

Provided that the protein and carbohydrate NMR signal has decayed and the moisture is dried out, the attenuation in Eq.1 is simplified to

$$I = I_{bound\_fat}\, e^{-\dfrac{2n\tau}{T_2^{bound\_fat}}} + I_{free\_fat}\, e^{-\dfrac{2n\tau}{T_2^{free\_fat}}} \tag{2}$$

Thus the lipid signal is resolved and total lipid content can be found by a two exponential fit of equation as a function of echo number **n**. Even though the experiment in principle is straight forward, the successful resolution of lipid signal is dependent on a careful setting of the experimental parameters to start the recording of the echo train as early as possible without any interference from other macromolecules exhibiting proton signal.

2.2   Resolving lipids in fresh systems

To resolve the lipid signal from the water signal in fresh samples, we add pulsed magnetic field gradients (PFG) to the CPMG experiment. Then we are able to resolve the lipid signal due to differences in molecular mobility as well as transverse relaxation times. Lipids have a very different mobility compared to moisture [5], while the protein signal with transverse relaxation time less than 1ms will not contribute when the first measuring point (n=1) in the combined PFG-CPMG experiment is at approximately 5 ms or more [5].

Assuming that the moisture signal is suppressed by the use of pulsed magnetic field gradients and that the protein signal is decayed due to short transverse relaxation times, the echo attenuation for the combined PFG-CPMG sequence in Figure 1 is written

$$I = I_0^{lipid}\, e^{-\dfrac{2n\tau}{T_2^{lipid}} - \dfrac{2n\tau^3}{3}\gamma^2 G_i^2 D^{lipid}}\, e^{-n\gamma^2 g^2 D^{lipid} \delta^2 (\tau - \dfrac{\delta}{3})} \tag{3}$$

Thus the lipid signal is resolved and the lipid content can be found by an exponential fit of equation as a function of echo number **n**.

3   EXPERIMENTAL

The ordinary spin echo experiments were performed on a Bruker Minispec pc 120, 0.47 Tesla with 18 mm H-probe where the inter echo spacing in the single spin echo experiment, τ, is set to 310 μs. The low resolution CPMG and combined PFG-CPMG experiments were performed on a Maran 0.55 Tesla equipped with 18 mm. gradient H-probe and access to 350 Gauss/cm. The CPMG method has been optimised on two Maran 0.55 Tesla systems with different RF power available; 25 W yielding a 90 degree RF pulse of 15 μs, and 300 W yielding a 90 degree RF pulse of 4.5 μs. With 25 W RF power, a duration of the 180 degree RF pulse of 30 μs gave an optimum setting of the τ-value to 55 μs. If the τ-value was kept at 40 μs, as is the optimum setting using 300 W of RF power, signal from protein and carbohydrates will interfere due to $T_{1\rho}$ effects.

The high-resolution CPMG spectra were recorded on a 14.1 Tesla Bruker Avance DRX system. In the CPMG pulse sequence we used a τ value of 2 ms and collected data at 16 different echo times ranging from 8 ms to 4 seconds, and the duration of the 90 degree pulse was 11 μs. The marine powders investigated at high resolution were dissolved in deuterated chloroform and analysed directly. TMS (Tetra Methyl Silane) was used as an internal chemical shift reference.

The samples investigated were various types of homogenised fish, fish feed, and various types of marine powders as cod, salmon, crab, etc. The determined lipid contents from the NMR results were compared with results achieved from well approved standard chemical extraction techniques as Soxhlet [7], Ethyl Acetate [8], and Acid Hydrolysis [9]

When using the combined PFG-CPMG technique, the fresh fish was homogenised before 3.5 gram of it was placed in the NMR tube. Numerous measurements from the same bath indicated that this was sufficient to achieve proper homogenisation. To suppress the moisture signal we made use of 1.5 ms gradient pulses of 0.2 T/m, and the first echo recorded was echo number 4. To avoid unwanted loss of echo signal due to convection caused by possible temperature gradients and/or knocking of the gradient coils when switching the current on and off, only even echoes were recorded and used when finding the initial lipid signal [10-11].

With respect to the ordinary spin echo and the optimised CPMG technique, the moisture was dried out prior to the NMR measurements. An electric oven operating at 104°C was used for this purpose.

4 RESULTS AND DISCUSSION

4.1 Comparison between the m-PFGSE, the single spin echo experiment and a chemical extraction method on fish samples

As the gradient strength is increased in the m-PFGSE experiment (see Figure 1 and Equation 1), the moisture signal is more attenuated than the lipids due to its much higher mobility. To record the lipid signal only, the gradient strength and the duration of the gradient pulse is then adjusted such that there is no NMR echo signal that originates from the moisture. The gradient strength is adjusted such that the moisture signal is attenuated down to a fraction of e^{-10} of the initial signal. As the mobility of the lipids is found to be 100 times slower than the mobility of the moisture, we can conclude that the moisture signal is suppressed to an insignificant when the lipid signal is attenuated down to a fraction of $e^{-0.1}$ of the initial lipid signal. The initial lipid signal is then found from fitting the intensity of the echo peaks containing lipid signal only, to a one exponential decay as given in equation 3. Table 1 shows the results achieved with the combined PFG-CPMG method and compared to standard chemical extraction method using Ethyl Acetate. The values given are average values and quite representative for the technique used.

Even though the results are identical within the experimental noise, which is found to be ± 0.2 in absolute units for both techniques, there seems to be a systematic difference between the results achieved with the PFG-CPMG technique and by the use of chemical extraction. Such an offset could be explained by a small amount of lipid remaining in the tissue from which the chemical solvents should have extracted the lipid completely. To confirm the existence of this offset, homogenised filet of salmon, from which the lipid should have been extracted by the use of chemical solvents [7-9], were analysed in a 14.1 Tesla high-resolution spectrometer. The results shown in Figure 2 confirm that there are significant amounts of lipid left. The spectrum originates from this tissue dissolved in deuterated chloroform, and shows the characteristic shifts of triglycerides. Typical signals are the glycerol proton signals at 5.26 ppm ($-CH_2-CH-CH_2$) and 4.29-4.14 ppm ($-CH_2-CH-CH_2-$) [12, 13]. Other typical lipid signals occur at 5.34 ppm and in the region 2.90-0.80 ppm [12]. Using the optimised CPMG method the amount of remaining lipid within the tissue is found to be of the order of 0.1 % and may thus explain the discrepancy between the PFG-CPMG method and chemical extraction methods.

Table 1   *Comparison between determined fat content (%) from the m-PFGSE method and a chemical extraction method [8]*

Fat content / %	m-PFGSE	Chemical extraction
Wild salmon	5.5	5.3
Bred salmon	11.1	10.9
Mackerel	30.0	29.8
Herring	17.2	16.9

4.2   Comparison between the optimised CPMG and chemical extraction on marine powders and fish feed

We find that the ordinary spin echo technique fails in determining the total lipid in dried marine products. This is due to the fact that when the moisture is dried out, the mobility of part of the lipid which is coupled to a solid matrix is significantly reduced. As most of phospholipids are found in the cell membranes, it leads to a reduction in their transverse relaxation times when the tissue is subjected drying.

In Figure 3 the initial CPMG attenuation from cod powder is shown together with the resulting T_2 distribution. The acquisition of the CPMG attenuation starts at 180 us to avoid any interference of signal from protein and other macromolecules. The signal consists of two main components which are easily resolved using an inverse laplace transform on the CPMG attenuation [14]. This approach for determining the total lipid content has been applied on 30 different samples of various marine products (powdered coalfish, crab, bluebeard, salmon and a selection of powdered white fish). For comparison Acid Hydrolysis (AH) was applied on the powdered systems as well. In Figure 4 the NMR results are plotted against the AH results for powdered systems with lipid content ranging from 3 to 19 %. The slope of the linear fit within this range is close to 1.00, indicating that no significant discrepancy is found between the two methods. However, if we had included the results from salmon powder where the lipid content varies from 16 % to 33 % we find that the NMR method systematically yields a higher lipid content than the by using the AH

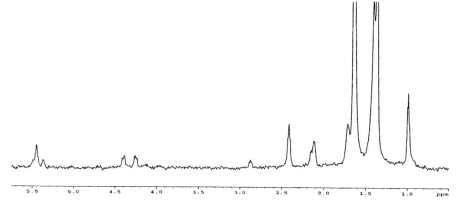

Figure 2   *Identification of lipids in chemical extracted tissue*

method. As is shown in Figure 2, the most likely cause of this discrepancy is that the NMR method measures the total lipid content, while the AH method does not necessarily extract all lipids from a sample with high lipid content. This is confirmed when a double chemical extraction is performed on a selection of marine powders. Then the measured lipid content by chemical extraction is increased and we do no longer find significant discrepancies between those and the results achieved with NMR method.

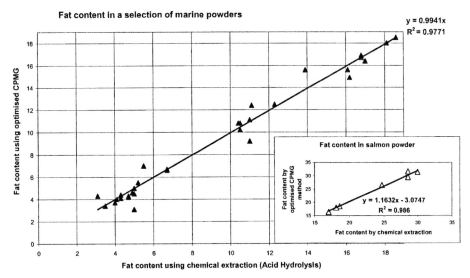

Figure 3 *CPMG attenuation from dried cod powder and the corresponding T2 distribution*

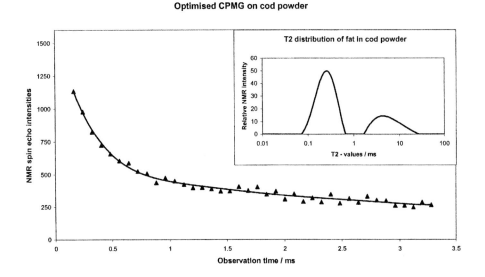

Figure 4   *Comparison between optimised CPMG method and chemical extraction by Acid Hydrolysis*

4.3   The resolution of phospholipids from other lipids using the optimised CPMG

As shown in Figure 3, the transverse relaxation times can be divided into two clearly separated regions for the lipids in the marine powders and other dried systems. The most obvious explanation to this is the different behaviour of phospholipids and other lipids when there are small or no amounts of moisture in the samples. The mobility of the phospholipids will be strongly reduced, because the phospholipids are attached to a much more rigid structure compared to a wet system where the moisture makes the texture less rigid. The other lipids will be less affected by the low moisture content as they still will be in a liquid state.

When analysing the high-resolution spectra from cod (upper spectrum) and salmon powder, it is evident that the dominant lipid in cod powder is phospholipids. This is particularly seen from the phospholipid head group signals at 4.42 ppm ($-O-CH_2-$) and 3.37 ppm ($[-CH_3]_3-N^+$) and from the glycerol signals at 5.25 ppm ($-CH_2-CH-CH_2$) and 4.44-3.78 ($-CH_2-CH-CH_2$) [13, 15]. From the cod powder spectrum, we also see that both unsaturated and saturated fatty acids are bound to the phospholipids. Unsaturated fatty acids give the following resonances: 5.44-5.28 ppm ($-HC=CH-$), 2.89-2.76 ppm ($=CH-CH_2-CH=$) and 2.07 ppm ($=CH-CH_2-$) (2, 4). The signal at 0.97 ppm, in addition to the usual $-CH_3$ signal at 0.87 ppm, indicates an n-3 unsaturated fatty acid [16-18]. n-3 unsaturated fatty acids, such as DHA (docosahexaenoic acid; C22:6(n-3)) are well known components in fish oils and are mostly attached to the *sn*-2 position of phospholipids [19].

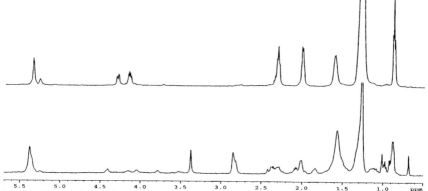

Figure 5   *High-resolution NMR spectra arising from Cod powder (lower spectrum) and Salmon powder (upper spectrum).*

The high-resolution CPMG spectra on the dissolved powders show that the peaks arising from phospholipids exhibit a much shorter T_2 than the peaks arising from the other lipids. This is in accordance with the results achieved on the low-field instrument. Here we find that the relative fraction of lipids with short T_2 is 0.75 of the total lipid content in the cod powder while it is only 0.05 in the salmon powder. This supports the picture of a low resolution NMR method that is capable of resolving the amount of phospholipids from other lipids.

5 CONCLUSION

The combined PFG-CPMG method [4] determines the lipid content in fresh samples of fish as accurate as extraction methods [7-9] and using the spin echo method on dried samples [2-3]. Comparison between the PFG-CPMG method and the ordinary spin echo method indicates that the ordinary spin echo method may have difficulties quantifying the phospholipids within the dried fish sample.

While both the PFG-CPMG and the spin echo methods fail to determine the total lipid content in powdered systems with small amounts of moisture, the optimised CPMG methods quantifies the total lipid content as accurate as the chemical extraction methods. There are also strong indications that the two-component behaviour in transverse relaxation time can be assigned to phospholipids ($T_2 \sim 0.2$ ms) and other lipids ($T_2 \sim 100$ ms).

Acknowledgements

We thank Toro Rieber & Sønn for the supplying of marine powdered samples and for letting us use the results from total lipid content measurements achieved with acid hydrolysis, and we thank Nutreco ARC for supplying us with chemically extracted tissue.

References

1. Resonance Instruments Ltd,31a Avenue One, Witney, Oxfordshire, OX28 4XZUK, AN026 2002;v.1.0
2. Collins M. J., King E. E.. *United States Patent* 6,548,303 , 2003
3. ISO 10565, 1998
4. Sørland G.H., Larsen P.M, Lundby F. Rudi A., Guiheneuf T.. *Meat Science* 2004;66, 543-550
5. Bottomley, P.A, Hardy, C.J., Argersinger, R.E & Allen-Moore, G. *Medical Physics* 1987;14(1)
6. Abragham A., "Principles of Nuclear Magnetism", *International series of monographs on physics-32*, Oxford University Press , 1961
7. AOAC 1997;991.36
8. NS(Norsk standard) 1994;9402
9. ISO 1973;1443
10. Callaghan P.T., Xia Y. *J. Magn. Reson.* 1991;91:326
11. Sørland G.H., Aksnes D.. *Magn. Reson. Chem.* 2002; 40:139-146
12. Haraldsson G, Gudmundsson B, Almarsson O..*Tetrahedron.* 1995; 51: 941-952.
13. Kriat M, Viondury J, Confortgouny S, Favre R, Viout P, Sciaky M, Sari H, Cozzone PJ.. *J. Lipid Res.* 1993; 34: 1009-1019.
14. Resonance Instruments Ltd, 31a Avenue One, Witney, Oxfordshire, OX28 4XZUK. *RI WinDXP* 1998;v. 1.3
15. Lammers JG, Liefkens TJ, Bus J, Vandermeer J.. *Chem. Phys. Lipids.* 1978; 22: 293-305.
16. Aursand M, Rainuzzo JR, Grasdalen H.. *J. Am. Oil Chem. Soc.* 1993; 70: 971-981.
17. Gunstone FD.. *Chem. Phys. Lip.* 1991; 59: 83-89.
18. Diehl BWK.. *Eur. J. Lipid Sci. Technol.* 2001; 103: 830-834.
19. Holte LL, van Kuijk FJGM, Dratz EA.. *Anal. Biochem.* 1990; 188: 136-141.

Authenticity and Quantification of Food

ADVANCES IN THE AUTHENTICATION OF FOOD BY SNIF-NMR

Gerard Martin

Professor Emeritus of Nantes University
Eurofins Scientific

1   ABSTRACT

Recent developments and new applications of SNIF-NMR in Food Science are reviewed under several topics: methodological improvements, alcoholic beverages, fruit juices, aromas and perfumes, fats and oils, milk, and drugs.

2   INTRODUCTION

Site-specific natural isotope fractionation studied by Nuclear Magnetic Resonance (SNIF-NMR) will celebrate next year its 25th birthday. Primarily devoted to the authentication of wines, many other applications succeeded in nearly all the fields of Food Science. Since SNIF-NMR was presented a decade ago, during a former Conference on the Applications of Magnetic Resonance in Food Science, the purpose of this contribution is to review the new achievements done from the beginning of the nineties until now.

3   METHODOLOGY

No spectacular improvement of the SNIF-NMR methodology itself occurred in the last ten years: a spectrometer working at 11.4 T, fitted with a specific ^{2}H$\{^1$H$\}$ 10 mm OD probe and a ^{19}F locking device represents a fairly optimal configuration. Obvious improvements such as higher field spectrometers and/or specific cryoprobes are rather a matter of financing than of originality. However, advances in the signal treatments and in the integration of hyphenated isotope ratio mass spectrometry in a SNIF-NMR strategy show a nice level of achievement. In fact, the disadvantage of the low intrinsic sensitivity of NMR is amplified by the small abundance of the heavy isotopes of the main elements (2H, 13C, 15N, 17O) of organic molecules, and increasing the signal-to-noise ratio is the main challenge to overcome [1].

2H	13C	15N	17O
V-SMOW	V-PDB	Air	V-SMOW
0.00015574	0.01111233	0.00366303	0.000401

Table 1 *Mean isotopic abundances of the heavy isotopes, having a magnetic moment, of organic molecules*[2,3]

Nevertheless, good precision and accuracy are obtained now by deuterium SNIF-NMR which compete favourably with mass spectrometry [4] (Table 2)

Analytical Technique	Product Concerned	Substrate Studied	Repeatability Sr (‰)	Reproducibility SR (‰)	Normalization Institute
MS	Juices	Water	1.3	2.7	ECN 1994
NMR	Wines	Ethanol	1.2	1.8	JRC-EU 1994
NMR	Juices	Sugar/Ethanol	1.2	1.8	ECN/AOAC 1995

Table 2 *Precision of SNIF-NMR as compared to IRMS for the determination of (D/H) isotope ratios. **ECN** : European Committee of Normalization, **AOAC** : Ass. Official Analytical Chemists, **JRC-EU** : Joint Research Centre-European Union.*

The isotope ratios may be obtained either with an internal reference or without any reference. The internal reference or working standard (ws) is usually a chemical (N,N-tetramethylurea, dioxane, hexamethyldisulfide) but it has been proposed recently to generate an electronic signal in the spectrum of the molecule (A) under study, the so-called ERETIC method [5]. The idea of ERETIC is to use as a reference, not a real NMR signal but rather a pseudo-FID, synthesized by an electronic device and transmitted inside the NMR probe during the reception of the NMR signal. If this electronic signal has the same shape and the same frequency as the FID, then it gives, after Fourier transform, an additional peak in the spectrum. The ERETIC signal is usually calibrated by proton NMR. In the case of the internal referencing procedure, we have:

$$Ri = \frac{p^{ws}\, m^{ws}\, M^{A}\, I^{A}\, R^{ws}}{p^{A}\, m^{A}\, M^{ws}\, I^{ws}} \qquad \text{(eq 1)}$$

$$Ri = \frac{fi\, R^{ms}}{Fi} \qquad \text{(eq 2)}$$

Ri, Rws are respectively the isotope ratios of isotopomer I and of the working standard, whatever the nucleus under study ($^2H/^1H$, $^{13}C/^{12}C$, $^{15}N/^{14}N$, $^{17}O/^{16}O$) and R ^{ms} is the isotope ratio of the whole molecule. R^{ms} is generally determined by isotope ratio mass spectrometry (IRMS). fi is the molar ratio of isotopomer I , computed from the intensities of the NMR signals and Fi, the statistical molar ratio is given by:

$$Fi = pi/\Sigma pi \qquad \text{(eq 3)}$$

Primarily devoted to the determination of (D/H) isotope ratios, SNIF-NMR has been extended to other nuclei, ^{13}C, ^{15}N and ^{17}O [2,3,6]. However, in the cases of nitrogen and oxygen, the acquisition conditions are far too much severe for being applied in routine analyses of foods and beverages.

4 APPLICATIONS TO FOOD SCIENCE

4.1 Alcoholic beverages

Frauds on wines are as old as the human being! In the Roman Empire, fine Latium wines were already blended with plonk. According to the Liverpool Mercury in 1854, "it was public knowledge that much more Port wine is drunk in the Kingdom than produced in Portugal". By a strange irony of fate, it seems that Engels became in 1844 the advocate of the official control laboratories when he stated in his book, "The Condition of the Working Class in England: as the industrial capitalism expanded, it appeared clearly that only fear of the law might restraint capitalists to sell adulterated or even toxic goods".

Nowadays, the more significant development in the analysis of wines is the implementation by the European Community of a large databank constituted of up to twelve thousands of isotopic profiles of Quality Wines Produced in Specific Regions (QWPSR)[7]. All the wine-producing countries of the EU (Austria, France, Germany, Greece, Italia, Portugal, Spain and even Luxembourg and UK) harvest every year a number of samples proportional to the area of their vineyards (500 for Italia, 450 for France, 220 for Germany and Spain). The databank contains the SNIF-NMR parameters of ethanol,(D/H)I, and the ^{13}C and ^{18}O IRMS data of ethanol and water respectively. The purpose of this databank is to provide the Official Laboratories of the Member States with reference data for checking the conformity of wines as regards to the current regulations. More recently, wines from third party or newly integrated countries, Bulgaria, Hungary, Slovenia, Croatia, were also subjected to specific studies. Fortified and enriched (chaptalised) wines are quantified by using two isotopic parameters (D/H)I and $\delta^{13}C$, which are obtained by SNIF-NMR and IRMS respectively. From the European databank, it is possible to construct ellipses defining the membership of natural wines from a given region and year for a confidence level of 95% or more. Figure 1 shows some examples of dubious Northern wines in the plan of (D/H)I vs $\delta^{13}C$: the 95% and 99% confidence level ellipses are constructed with the mean data of the population of natural wines from a Northern region ((D/H)I= 101.0 ppm and $\delta^{13}C$= -27.0 ‰). It is apparent from the consideration of Figure 1 that some wines may be found in blurred regions of the plan and the analyst should give his judgment in terms of first or second order risk. Recently, isotopic analysis of glycerol contained in wines was applied to precise the conditions of deuterium transfer from sugar to the fermentation products [6,8].

The authentication of spirits is generally carried out in the same ways as for wines. Pure grain vodkas may be contaminated with potato or beet ethanol, tequila and sake are corrupted by cane ethanol, brandies are substituted to cognac, pure malt whiskies are blended with grain ethanol.

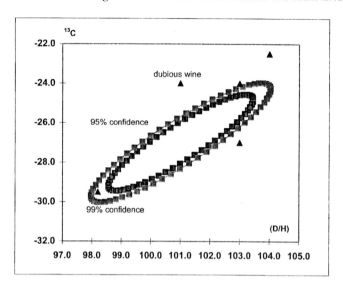

Figure 1 *Bivariate representation of European wines in the plan of (D/H)I and ^{13}C*

4.2 Fruit juices

At the beginning of the nineties, the market of orange juices was in a so bad condition in terms of quality and economic competition that enforcement regulation organisms got huffy in the USA and the UK. The MAFF at this time organised a large campaign of control of juices taken from the supermarkets. The frauds on juices are manifold: sucrose added to poor strength juices, pure juices made in fact from concentrates and diluted, synthetic organic acids added to adjust the pH and, that is more precisely an economic fraud, juices from country A, stricken by high custom taxes, imported under the banner of country B, enjoying favoured custom regulations. The addition of beet or cane sucrose to orange juices is determined in the same manner as for wines: In the plan of (D/H)I vs $\delta^{13}C$, anomalous samples are readily identified and the quantity of sucrose added is computed accordingly. This procedure was made possible after a series of studies concerning the isotope transfer from glucose into ethanol during fermentation. The fractionation factors, measured in 1986 [9] and refined some years later [10], gives good means to relate unambiguously the couples glucose-water and ethanol-water. Typically, the isotopic parameters of three important fruit juices are shown in Table 3.

Juice	Water		Ethanol		
	(D/H)ws	δ18Ows	(D/H)I	(D/H)II	δ13C
Apple	151.6	-1.9	97.8	125.2	-26.7
Pineapple	154.8	2.5	110	132.2	-14.7
Orange	156.5	3.0	105.1	129.6	-26.7

Table 3 *Main isotopic parameters of the three most consumed fruit juices. D/H)ws (ppm) and δ18Ows(‰) are the parameters of the water of the juices before fermentation and (D/H)I, (D/H)II, (ppm), δ13C(‰) are the parameters of ethanol after fermentation.*

Since the fraud consisting to add sucrose to the juices seems to be now nearly eradicated (at least at the 5% level!), the distinction between pure natural juices and juices made from concentrate remains a problem. In fact, fraudsters have found a nice way to bypass the accepted isotopic method based on the determination of the parameters of water of juices. At least in the case of orange, the tap water used to dilute a concentrate has much lower isotopic values, on the order on 150 ppm et –6.5‰, than those of water of juices (Table 3) and the conventional fraud can be detected.

It has been noted that large quantities of water resulting from some kind of processes, i.e. must concentration in Europe, are available at nearly free of charge. Since this water has higher isotopic values than that of tap water, it is possible to mime the isotopic properties of a natural juice by diluting a concentrate by the processed water. In order to fight against this practise, Eurofins Scientific has developed an exhaustive data bank on orange juices from the main producing countries in the world (Brazil, Florida, Mexico, Belize, North and South Africa, Israel, Spain, Italia...) The data bank gives confidence limits for the juices of a given country and a specific procedure, based on the thermodynamic and kinetic fractionation factors of the evaporation processes of juices, is able to identify an adulterated juice in a number of cases. More over, getting IRMS into a global authentication strategy [11] increases significantly the deterrent power of isotopic analysis in fighting frauds in juices.

4.3 Aromas and Perfumes

Vanillin is very good example of the continuous fight between Ariel and Caliban! Since natural vanillin extracted from beans costs about one hundred times more than synthetic vanillin from lignin or guaiacol, the leading objective of fraudsters was to mime the isotopic properties of natural vanillin with a manipulated synthetic product. Short history of the fight:

< 1980 vanillin ex lignin is frequently sold in place of vanilla beans

1980: ^{13}C IRMS forces back fraudsters [12].

≈1983-1985 ^{13}C IRMS is defeated by addition of labelled vanillin in vanillin ex lignin [13,14].

1988 ^{2}H-SNIF-NMR forces back fraudsters [15].

»1990 Skilled chemists attacked ^{2}H-SNIF-NMR by exchanging some typical sites of vanillin by heavy water.

1996 European regulations consider as fully natural biotechnological vanillin and give happy life to the fraudsters since no method exists to authenticate biotechnological vanillin [16].

1996-1997 ^{2}H-SNIF-NMR authentication of both vanillin and p hydroxy benzaldehyde in vanilla beans makes the fraudsters in uncomfortable position [17,18].

2002-2003 A joint procedure of isotopic analysis by NMR and trace analysis by HPLC is able to characterize biotechnological vanillin obtained from ferulic acid [19,20]

The three main groups of synthetic vanillin are obtained from guaiacol (o-methoxyphenol) and the formyl group results from an electrophylic addition of formaldehyde or glyoxylic acid. As a consequence of more or less important kinetic isotope effects, the deuterium content of the CHO fragment is much greater than that of the natural products, bur for vanillin ex eugenol. Mechanistic studies [18] have demonstrated the nature of the enrichment of the formyl group. A statistical computation on the deuterium content of the aromatic sites 5 indicates that this parameter is a very efficient authenticity criterion at the 99% confidence level for vanillin ex *Vanilla planifolia*. It is also interesting to note that the isotope ratios of the methoxyl group of the three fossil vanillin samples are on the order of

138 ppm against 126 ppm for the vanillin samples from a living source, in agreement with the data observed for fossil and natural raw materials. The very low value of $(D/H)_{OCH3}$ for vanillin ex lignin is related to the oxidative degradation of the phenylpropanoid chain of lignin. To conclude the examination of Table 4, it is worth to discuss the two extreme values of $\delta^{13}C$: the lower isotopic deviation is a good proof for the natural status of ferulic acid and the higher value, characteristic of a CAM photosynthetic metabolism unfortunately favours labelling manipulation of vanillin, since it is much more easy to add rather to remove heavy isotopes in a product.

Group	Fossil vanillin			Semi Synth.	Natural vanillin		
	Formaldehyde	Glyoxilic a. 1	Glyoxilic a. 2	Lignin	Eugenol	Ferulic a.	Vanilla p.
$(D/H)_{CHO}$	204.0	240.0	358.7	121.5	194.7	124.3	131.1
(ppm)	12.6	13.3	23.9	4.5	12.6	2.9	3.0
$(D/H)_{Ar5}$	145.1	138.5	141.1	168.5	168.1	146.1	197.1
(ppm)	4.8	6.4	6.1	4.5	7.3	3.4	3.3
$(D/H)_{OCH3}$	137.1	141.4	136.4	105.5	125.8	127.1	127.2
(ppm)	8.8	10.2	4.1	1.3	1.9	2.5	1.6
$\delta^{13}C$	-30.5	-26.8	-28.3	-26.7	-30.0	-36.2	-20.2
(‰)	0.8	1.1	0.7	0.5	0.8	0.3	0.2

Table 4 *Isotopic parameters of the main groups of vanillin available in the food industry*

The analytical methods were strengthened by studying para hydroxybenzaldehyde (pHB) usually present in a small quantity ($\approx$10%) in vanilla beans. Using a similar approach as for vanillin, the three $(D/H)_i$ ratios of pHB enable to distinguish between naturally occurring and synthetic species since, in manipulated vanilla aromas, synthetic pHB is mixed to synthetic vanillin.

A comparison of the ^2H isotopic data of pHB and vanillin from beans at the same position reveals that the aromatic ring shows identical values. However, the deuterium content of the formyl group is very different in both compounds (Vanilla: 130 ppm and pHB: 240 ppm). Since previous studies have concluded that both vanillin and pHB in beans are biosynthesised in the same way, the difference of ^2H abundance of the aldehyde group is due to kinetic isotope effect on the last step of the metabolism. The isotopic balance demands that the ^2H depletion noticed for vanillin, as the major product, is compensated by ^2H enrichment for pHB being the minor product [17]

Other flavouring molecules are authenticated by SNIF-NMR. For example, the mains active molecules in bitter almond and cinnamon oils are benzaldehyde and cinnamaldehyde, respectively. The market is divided into four main sources of benzaldehyde: (i) direct oxidation of toluene; (ii) hydrolysis of benzal chloride, produced itself by chlorination of toluene; (iii) retroaldolization of cinnamadehyde; and (iv) hydrolysis of amygdalin from bitter almond and kernels (apricots, peaches, plums, cherries. Cinnamaldehyde may be either synthetic or extracted from natural material (cinnamon oil or cassia oil) and in that case benzaldehyde is classified as semi-synthetic. Interestingly, synthetic cinnamaldehyde is issued from benzaldehyde and, therefore a close relationship exists between these two molecules. This fact was applied [21] to the study of origin of cinnamaldehyde by its transformation into benzaldehyde, since the ^2H-NMR spectrum of cinnamaldehyde exhibits unfavourable signal overlaps, even at a high magnetic field. Thus, benzaldehyde is a common molecular probe for studying the authenticity of both bitter almond and cinnamon oils. The reliability of the method includes the chemical transformation of cinnamaldehyde to benzaldehyde, performed at the laboratory scale. The SNIF-NMR method has been applied to allyl isothiocyanate in order to authenticate mustard oil samples [22]. Prior to this work no analytical methods were available.

Recently, other aromatic benzenic molecules were studied by a Milanese group: phenylethanol and phenylethyl acetate [23] and raspberry ketone [24,25]

The importance of monoterpenes in the field of aromas and perfumes but also in plant physiology created also a strong interest for SNIF-NMR studies. For example, Hanneguelle et al [26] studied carefully the case of linalool. A number of samples were extracted from plants and obtained by semi synthetic and synthetic ways in order to define authentication criteria. More recently [27], the isotopic connectivity between 25 terpenic molecules was described at the light of the new model of monoterpene biosynthesis, the deoxyxylulose pathway (DOXP).

4.4 Miscellaneous recent applications of SNIF-NMR

Fats and oils [28,29], fishes [30,31], milk and dairy products [32], coffees[33] received a great attention form the official and private laboratories in charge of the consumer protection. Legal, tobacco[34,35], and illegal, heroin[36] , drugs have also been authenticated by SNIF-NMR.

For example, olive oil was studied with a view of checking its adulteration with cheaper oils such as hazelnut oil. For this purpose, a thorough investigation of the European sources of olives (region and year of production) and of the characteristics of production (maturity of olives, composition of the paste, technology) was undertaken in 1997-1999. The SNIF-NMR methodology was adapted to the specific case of fatty acids, which display much less isotopomers in the ^{2}H-NMR spectra than the number of non-equivalent sites. Then, the concept of isotopic clusters, gathering several overlapping signals, replaced that of individual isotopic signal. Moreover, since it is nearly impracticable to quantitatively isolate pure fatty acids after hydrolysis of triglycerides, the hydrolysis mixture was studied as a whole and a very careful GC determination of its composition was required.

The authentication of wild salmons against farmed fishes is based on the same methodology since the fat of fishes is a better origin marker than proteins. It was possible to identify farmed salmons from Scotland and Norway and distinguish them from wild salmons caught in the Atlantic or Pacific oceans.

Lactose, fermented into ethanol and triglycerides are sensitive probe to relate the isotopic properties of the animal diet to those of milk. In the case of coffee, the study of caffeine gives indications about the natural or synthetic status of the product and enables to differentiate *Arabica* from *Robusta* plants. Finally, although tobacco and heroin are not really foodstuffs, unfortunately they are consumable goods and must be identified.

References

1   G. J. Martin, M. L. Martin, Y. L. Martin *Bull.Magn.Reson.* 1996, 44075.
2   G. J. Martin *Git Labor-Fachzeitschrift* 1998, 494-498.
3   V. Caer, M. Trierweiler, G. J. Martin, M. L. Martin *Anal. Chem.* 1991, 2306-2313.
4   B. L. Zhang, M. Trierweiler, C. Jouitteau, G. J. Martin *Anal. Chem.* 1999, 2301-2306.
5   I. Billault, R. Robins, S. Akoka *Anal.Chem.* 2002, 5902-5906.

6   B. L. Zhang, S. Buddrus, M. Trierweiler, G. J. Martin *J.Agric.Food Chem.* 1998,
     1374- 1380.
7   Commission Des Communautés Européennes *J. Officiel des Communautés
     Européennes*, 1991, 39-43.
8   B. L. Zhang, Yunianta, V. Bordage, S. Buddrus, G. J. Martin *J. Agric. Food Chem.*
     1998.
9   G. J. Martin, B. L. Zhang, N. Naulet, M. L. Martin *J. Am. Chem. Soc.* 1986, 5116-
     5122.
10   B. L. Zhang, Yunianta, M. L. Martin *J.Biol.Chem.* 1995, 16023-16029.
11   E. Jamin, R. Guerin, M. Retif, L. Lees, G. J. Martin *J. Agric. Food Chem.* 2003, 5202-
     5206.
12   J. Bricout, J. Koziet *Labo-Pharma Probl.Tech.* 1980, 490.
13   D. A. Krueger, H. W. Krueger *J.Agric.Food Chem.* 1983, 1265-1268.
14   D. A. Krueger, H. W. Krueger *J. Agric. Food Chem.* 1985, 323-325.
15   C. Maubert, C. Guerin, F. Mabon, G. J. Martin *Analysis* 1988, 434-439.
16   Commission Des Communautés Européennes *Joce* 1996, 1-4
17   G. S. Remaud, Y. L. Martin, G. G. Martin, G. J. Martin *J. Agric. Food Chem.* 1997,
     859-866.
18   G. J. Martin *Industrial Chemistry Library, Ed.J.R. Desmurs and Serge Ratton* 1996,
     506-527.
19   F. F. Bensaid, K. Wietzerbin, G. J. Martin *J. Agric. Food Chem.* 2002, 6271-6275.
20   F. Bensaid *Thèse Institut National Agronomique, Paris* 2003.
21   G. Remaud, A. A. Debon, Y. L. Martin, G. G. Martin, G. J. Martin *J. Agric. Food
     Chem.* 1997, 4042-4048.
22   G. S. Remaud, Y. L. Martin, G. G. Martin, N. Naulet, G. J. Martin *J. Agric. Food
     Chem.* 1997, 1844-1848.
23   G. Fronza, C. Fuganti, P. Grasselli, S. Servi, Zucchi *J. Agric. Food Chem.* 1995, 439-
     443.
24   G. Fronza, C. Fuganti, C. Guillou, F. Reniero, P. Joulain *J. Agric. Food Chem.* 1998,
     248- 254.
25   G. Fronza, C. Fuganti, G. Pedrocchi-Fantoni, S. Serra, Z. G., C. Fauhl, C. Guillou, F.
     Reniero *J. Agric. Food Chem.* 1999, 1150-1155.
26   S. Hanneguelle, J. N. Thibault, N. Naulet, G. J. Martin *J. Agr. Food Chem.* 1992, 81-
     87.
27   G. J. Martin, S. Lavoine-Hanneguelle, F. Mabon, M. L. Martin *Phytochemistry* 2004.
28   A. Royer, C. Gerard, N. Naulet, M. Lees, G. J. Martin *J. Am. Oil Chem. Soc.* 1999,
     357-363.
29   A. Royer, N. Naulet, F. Mabon, M. Lees, G. J. Martin *J. Am. Oil Chem. Soc.* 1999,
     365-373.
30   M. Aursand, F. Mabon, G. J. Martin *Jaocs* 2000, 659-666.
31   M. Aursand, F. Mabon, G. J. Martin *Magn.Reson.Chem.* 1997, S91-S100.
32   C. Vallet, Z. Mas'ud, G. J. Martin *J.Agric.Food Chem.* 1999, 4693-4699.
33   D. Danho, N. Naulet, G. J. Martin *Analusis* 1992, 179-184.
34   E. Jamin, N. Naulet, G. J. Martin *Photochem.Anal.* 1997, 105-109.
35   E. Jamin, N. Naulet, G. J. Martin *Plant Cell Environm.* 1997, 589-599.
36   P. Hays, G. Remaud, E. Jamin, Y. L. Martin *Journal of Forensic Sciences* 2000, 552-
     562.

HPLC-SPE-NMR: A PRODUCTIVITY TOOL FOR DETERMINATION OF NATURAL PRODUCTS IN PLANT MATERIAL

Jerzy W. Jaroszewski

Department of Medicinal Chemistry, The Danish University of Pharmaceutical Sciences, Universitetsparken 2, DK-2100 Copenhagen, Denmark. E-mail: jj@dfuni.dk.

1 INTRODUCTION

The presence of low-molecular weight natural products, i.e., the so-called secondary metabolites, is of crucial importance for the quality and value of foods, foodstuffs, nutritional supplements, spices, phytopharmaceuticals and other products of natural origin.[1-6] Secondary metabolites determine taste, odour and nutritional value of food. In some cases, they are believed to be instrumental for the maintenance of health. On the other hand, the presence of specific natural products can result in diminished quality of food or can be directly harmful, as in the case of natural toxins, unwanted products of degradation of natural food constituents, or contaminants resulting from adulteration, microbial infections, or pest control. For these reasons, techniques enabling identification and quantification of natural products are fundamentally important in food science.

Structural elucidation and quantification of natural products and other food constituents can be performed conveniently using hyphenated techniques, i.e., techniques that combine chromatographic separation with on-line spectroscopic characterization. In principle, such techniques enable simultaneous identification and quantification of numerous components of a complex mixture in one working procedure. A highly sophisticated hyphenated technique, albeit restricted to analysis of volatile constituents, is GC-MS, especially when coupled with a computerized database of mass spectra. Non-volatile constituents have to be separated using liquid-chromatographic techniques, and thus HPLC-DAD (spectrophotometric diode-array detection) and HPLC-MS (utilizing various mass-spectrometric techniques) have gained growing popularity over the years. However, UV-VIS provides very limited structural information and the same applies in part to HPLC-MS. Thus although the information about molecular mass of an analyte is highly valuable and additional useful information is provided when MS/MS techniques are included, the possibilities of differentiation between positional or geometrical isomers using mass spectrometry remain highly restricted.

Although NMR spectroscopy is considerably less sensitive than UV and MS, it constitutes the most general and informative spectroscopic technique available for organic compounds. Even a simple one-dimensional ^{1}H NMR spectrum recorded at a sufficiently high field can be extremely informative. For this reason, the idea of hyphenation of liquid-chromatography and NMR spectroscopy arose already before 1980.[7-12]

2 HPLC-NMR TECHNIQUES

Acquisition of NMR spectra of compounds eluted from liquid-chromatographic columns poses considerable challenges. In addition to low sensitivity, NMR spectra obtained in the HPLC-NMR mode have until recently been acquired in the same solvent and at the same concentration as provided by the chromatographic system. This creates a number of problems. Firstly, a solvent peak suppression sequence has to be included in the NMR experiment so that the dynamic range of receiver is not exceeded. Although effective solvent suppression schemes are now available, the reduction of solvent peaks is normally achieved on the expense of base-line distortions and intensity reduction of neighbouring resonances. Especially, effective suppression of broad water resonances poses considerable challenges. Therefore, many HPLC-NMR experiments reported in the literature have been performed using deuterated solvents in the HPLC mobile phases, with a penalty of cost and flexibility.

Traditional HPLC-NMR experiments are normally performed[13-15] in one of the following modes:

a) Continuous-flow HPLC-NMR. In this most simple version, NMR spectra are being acquired continuously with the HPLC column eluate. Since the eluate quickly enters and leaves the NMR flow cell, only one-dimensional ^{1}H NMR data can be obtained.

b) Stop-flow HPLC-NMR. In order to gain more time for the acquisition of NMR data, the solvent flow may be stopped during column elution. The halting of the elution may be done either in pre-defined time intervals (sliced stop-flow), or when a chromatographic peak reaches the NMR flow cell. Stop-flow experiments are in principle compatible with acquisition of time-consuming 2D NMR data. However, stopped flow causes diffusion-mediated broadening of chromatographic bands on the column, which can seriously compromise the experiment.

c) Peak-storage HPLC-NMR. The problem with peak broadening in stop-flow HPLC-NMR can be circumvented by complete elution of all compounds from the HPLC column into a loop-storage device. In this mode, multiple peaks can be stored practically indefinitely and subsequently transferred into the NMR flow probe for data acquisition.

3 HPLC-SPE-NMR

In the above-mentioned HPLC-NMR experiments, the sensitivity of data acquisition is determined by concentration of the analyte in the chromatographic elution band. Only a fraction of the peak elution volume that fits into the NMR flow cell contributes to the NMR signal, and the data are acquired in the HPLC solvent. The solvent choice is thus a compromise, because the optimal solvent for a particular HPLC separation may not be optimal for NMR and *vice versa*.

These limitations disappear by introduction of a solid-phase extraction (SPE) step between HPLC and NMR in order to remove the chromatographic solvent and replace it with a solvent more suitable for NMR spectroscopy. In this novel approach, solid-phase extraction cartridges are used to trap the analyte from the HPLC eluate. The SPE cartridge is then dried and the compound is eluted with an appropriate solvent into NMR flow cell.[16,17] The fully automated[17] post-column SPE process coupled with NMR is a substantial advance as compared to use of on-line SPE to concentrate the analyte prior to chromatographic separation in HPLC-NMR mode,[18,19] or as compared to SPE-NMR experiments using fractions from off-line HPLC separations.[20,21]

The first commercial HPLC-SPE-NMR system available includes a Spark Prospekt 2 solid-phase extraction unit (Spark, Emmen, Holland) (Figure 1) integrated with Bruker BioSpin HPLC-NMR system and operated with HyStar software (Bruker BioSpin, Karlsruhe, Germany). The system includes two trays (2 × 96 pieces) of 2 × 10 mm solid-phase extraction cartridges, which is more than sufficient for practical use. Many SPE phases are available, but materials such as C18 silica or polystyrene-type resin are obvious choices in conjunction with reverse-phase HPLC separations.

Figure 1   *Spark Prospekt 2 SPE unit carrying two trays with 96 SPE cartridges each*

One major advantage of HPLC-NMR-SPE is the possibility of multiple trappings on the SPE cartridges, which increases the amount of the analyte available for NMR experiments. Examples of use of this technique for structure determination of natural products are given below.

4   APPLICATION OF HPLC-SPE-NMR TO NATURAL PRODUCT IDENTIFICATION

4.1   Diterpenoids

Roots of *Perovskia* species (Lamiaceae) contain diterpenoid quinones – so called tanshinones – which are otherwise known as the active constituents of the famous traditional Chinese drug dan-shen.[22] Isolation of tanshinones from plant material is rather difficult and laborious, in part because plants contain series of closely related compounds, and in part because of the relative instability of tanshinones. HPLC-SPE-NMR experiments with crude ethyl acetate extracts of roots of *Perovskia atriplicifolia* Kar. rapidly provided 1D and 2D NMR data enabling elucidation of structures of all constituents of the extract, including compound **1** and cryptotanshinone (**2**) (Figures 2-4).

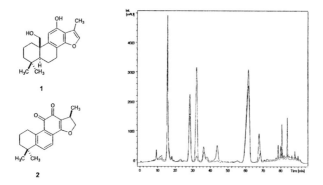

Figure 2   *HPLC (254 and 270 nm traces) of crude ethyl acetate extract of Perovskia atriplicifolia root (1.1 mg crude extract injected into Luna C18 column, 3 μm, 150 × 4.6 mm, acetonitrile gradient in water, 0.8 mL/min). For SPE trapping, the HPLC flow was diluted with 1 mL/min of water. Compound 1 and 2 (cryptotanshinone) have t_R = 32 min and t_R = 62 min, respectively.*

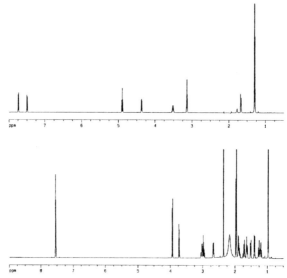

Figure 3   *600 MHz ¹H NMR spectra obtained in HPLC-SPE-NMR mode for peak with t_R = 32 (bottom, compound 1) and t_R = 62 min (top, cryptotanshinone 2). Bottom spectrum was obtained using 30 μL flow probe without solvent suppression by accumulation of 256 transients after six trapping on a 2 × 10 mm C18HD cartridge. Top spectrum was obtained with solvent suppression (1D NOESY) of the residual water and acetonitrile signals, accumulating of 256 transients after three trappings.*

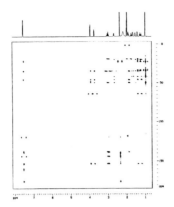

Figure 4   *600 MHz HMBC spectrum obtained in HPLC-SPE-NMR mode for peak with t_R = 32 min (compound 1). The spectrum was obtained using 30 µL flow probe following six trapping on a 2 × 10 mm C18HD cartridge.*

4.2   Flavonoids, isoflavonoids and flavonoid glycosides

Flavonoids are ubiquitous constituents of food plants. HPLC-SPE-NMR was used for identification of constituents of *Smirnovia iranica* H. Sabeti (Leguminoseae-Papilionoideae), a species related to *Glycyrrhiza*, which yields the licorice root. The latter genus is characterized by the presence of isoflavonoids,[23,24] and this type of natural products was also found in *Smirnovia iranica.*[25] HPLC-SPE-NMR proved to be a convenient method of structure elucidation of these constituents. Use of multiple peak trappings allowed acquisition of good quality HSQC and HMBC spectra, essential for unambiguous structure elucidation (data not shown).

Flavonoids are often present as flavonoid glycosides. In spite of increased polarity due to the presence of sugar residues, the glucosides are effectively trapped on SPE cartridges packed with octadecylsilyl silica. This was demonstrated by HPLC-SPE-NMR experiments with flavonoid glycosides from *Kanahia laniflora* (Forssk.) R. Br. (Apocynaceae), an African medicinal plant (data not shown).

4.3   Alkaloids

HPLC of alkaloids poses problems due to their basic character. Thus, reverse-phase chromatographic separations of alkaloids are normally preformed with acids or buffers in the mobile phase. Nevertheless, even dibasic alkaloids are well trapped on polystyrene-type SPE cartridges (GP phase), affording good quality NMR spectra. The example (Figure 5 and 6) shows separation of alkaloids from *Remijia peruviana* Standl., a species closely related to the quinine-producing *Cinchona* genus.

4.4   Cardioactive glycosides

Structure determination of the cardiac glycoside shown in Figure 7 is an example of HPLC-SPE-NMR experiment with a rather complex molecule.

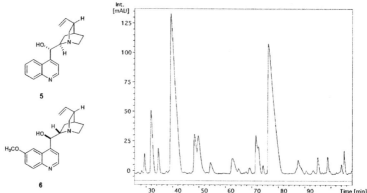

Figure 5   *HPLC (315 nm trace) of crude extract of Remijia peruviana bark (CH₂Cl₂ alkaline Soxhlet extraction); 100 μg of crude extract injected into Luna C18 column, 3 μm, 150 × 4.6 mm, acetonitrile gradient in water (0.08% trifluoroacetic acid), 0.8 mL/min. For SPE trapping (GP phase), the HPLC flow was diluted with 1 mL/min of water. Compound 5 (cinchonine) and 6 (quinine) have t_R = 39 min and t_R = 86 min, respectively.*

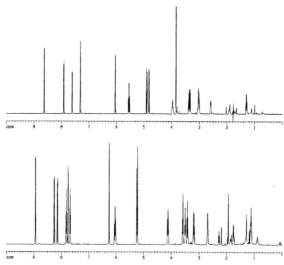

Figure 6   *600 MHz ¹H NMR spectra of 5 (bottom) and 6 (top) obtained in HPLC-SPE-NMR mode (30 μL cell) from the chromatogram shown in Figure 5. The spectra were obtained with solvent suppression after five trappings on 2 × 10 mm polystyrene-type GP SPE cartridges.*

4.5   Multiple trappings

Since multiple trappings are essential for HPLC-SPE-NMR experiments, it is of interest to explore trapping efficiency after repeated HPLC-SPE experiments. Figure 8 shows that a linear increase of the signal-to-noise ratio can indeed be obtained after multiple trappings, but choice of SPE material can be crucial for the outcome of the experiments.

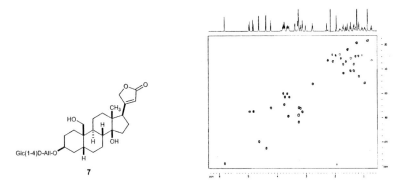

Figure 7 *600 MHz HSQC spectrum obtained with compound 7 in HPLC-SPE-NMR mode (six trapping on a 2 × 10 mm C18HD cartridge, 30 µL NMR flow cell)*

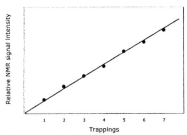

Figure 8 *Trapping efficiency of a flavonoid on octadecylsilyl silica*

5 CONCLUSIONS

The results described in this work demonstrate the broad applicability of HPLC-SPE-NMR with reverse-phase HPLC for studies of natural products. Examples including various classes of secondary metabolites (diterpenoids, flavonoids/isoflavonoids, flavonoid glycosides, cardenolides, quinoline alkaloids) were provided.

The major advantages of HLC-SPE-NMR are: a) solvent change, enabling the optimal solvent choice for HPLC and use of a deuterated solvent for NMR (since the NMR spectra are acquired in a well-defined solvent, unlike in traditional HPLC-NMR experiments, where the spectra are obtained in a mixture of solvents resulting from gradient chromatography, comparison with already available NMR data is facilitated); b) concentration of the whole HPLC elution band in a small (30 µL) NMR flow cell (unlike in traditional HPLC-NMR schemes, good results can be obtained with broad and asymmetric chromatographic peaks); c) possibility of multiple peak trappings, which increases the amount of material available for the NMR experiment. The latter is an important advantage enabling sensitivity increase at least equal to that attainable by use of cryogenic probes.[17,26] Because of the increased concentration of the analyte and use of deuterated solvents, NMR data can usually be acquired without solvent suppression. This yields spectra without base-line and intensity distortions and provides increased flexibility in NMR data acquisition in terms of optimization of read-pulse lengths and inter-pulse delays. Further developments in HPLC-SPE-NMR will comprise, inter alia, use of other chromatographic principles, use of novel SPE stationary phases, and computerised spectra interpretation. Work along these lines is in progress in our laboratories.

Acknowledgements

I am indebted to my friends, colleagues and students, in particular to C. Clarkson, M. Labmert, A. Maciuk, M. Sairafianpour, A.M. Rüsz and D. Stærk, whose work is mentioned in this article.

References

1   M. Iriti and F. Faoro, *Curr. Topics Nutraceut. Res.*, 2004, **2**, 47.
2   K. Brandt, L.P. Christensen, J. Hansen-Møller, S.L. Hansen, J. Haraldsdottir, L. Jespersen, S. Purup, A. Kharazmi, V. Barkholt, H. Frøkjær and M. Kobæk-Larsen. *Trends Food Sci. Technol.*, 2004, **15**, 384.
3   P.K. Lai and J. Roy, *Curr. Med. Chem.*, 2004, **11**, 1451.
4   R.H. Liu, *Appl. Biotechnol.*, 2003, **1**, 39.
5   W. Pfannhauser, G.R. Fenwick and S. Khokhar, eds., *Biologically Actiove Phytochemicals in Food*, Royal Society of Chemistry, Cambridge, 2001.
6   F. Mellon, R. Self and J.R. Startin, *Mass Spectrometry of Natural Substances in Food*, Royal Society of Chemistry, Cambridge, 2000.
7   N. Watanabe and E. Niki, *Proc. Jpn. Acad. Ser. B*, 1978, **54**, 194.
8   N. Watanabe, E. Niki and S. Shimizu, *JEOL News*, 1979, **15A**, 2.
9   E. Bayer, K. Albert, M. Nieder, E. Grom and T. Keller, *J. Chromatogr.*, 1979, **186**, 497.
10 E. Bayer, K. Albert, M. Nieder, E. Grom and Z. An, *Fresenius Zeitschr. Anal. Chem.*, 1980, **304**, 111.
11 E. Bayer, K. Albert, M. Nieder and E. Grom, *Anal. Chem.*, 1982, **54**, 1747.
12 D.A. Laude Jr. and C.A. Wilkins, *Anal. Chem.*, 1984, **56**, 2471.
13 K. Albert, ed., *On-Line LC-NMR and Related Techniques*, John Wiley & Sons, Chichester, 2002.
14 J.C. Lindon, N.J. Bailey, J.K. Nicholson and I.D. Wilson, *Handbook Anal. Sep.*, 2003, **4**, 293.
15 W. Kraus, L.H. Ngoc, J. Conrad, I. Klaiber, S. Reeb and B. Vogler, *Phytochem. Rev.*, 2003, **1**, 409.
16 L. Griffiths and R. Horton, *Magn. Res. Chem.*, 1998, **36**, 104.
17 V. Exarchou, M. Godejohann, T.A. van Beek, I.P. Gerothanassis and J. Vervoort, *Anal. Chem.*, 2003, **75**, 6288.
18 M Godejohann, A. Preiss, C. Mügge and G. Wünsch, *Anal. Chem.*, 1997, **69**, 3832.
19 J.A. de Koning, A.C. Hogenboom, T. lacker, S. Strohschein, K. Albert and U.A.Th. Brinkman, *J. Chromatogr.*, 1998, **813**, 55.
20 N.T. Nyberg, H. Baumann and L. Kenne, *Magn. Res. Chem.*, 2001, **39**, 236.
21 N.T. Nyberg, H. Baumann and L. Kenne, *Anal. Chem.*, 2003, **75**, 268.
22 M. Sairafianpour, J. Christensen, D. Stærk, B. A. Budnik, A. Kharazmi, K. Bagherzadeh and J. W. Jaroszewski, *J. Nat. Prod.*, 2001, **64**, 1398.
23 T. Nomura, T. Fukai, T. Akiyama, *Pure Appl. Chem.*, 2002, **74**, 1199.
24 T. Nomura and T. Fukai, *Progr. Chem. Org. Nat. Prod.* 1998, **73**, 1.
25 M. Sairafianpour, O. Kayser, J. Christensen, M. Asfa, M. Witt, D. Stærk and J. W. Jaroszewski, *J. Nat. Prod.* 2002, **65**, 1754.
26 M. Spraul, A.S. Freund, R.E. Nast, R.S. Withers, W.E. Maas and O. Corcoran, *Anal. Chem.*, 2003, **75**, 1541.

APPLICATIONS OF HYPHENATED NMR TO THR STUDY OF FOOD

M. Spraul, E. Humpfer and H. Schäfer

Bruker BioSpin, Rheinstetten Germany

1 INTRODUCTION

NMR-Hyphenation with high pressure liquid chromatography (LC-NMR) is a well established tool in many analytical laboratories. It has gained further momentum by the rapidly growing interest in metabolic profiling over the last 2 years, known under the keywords Metabonomics and Metabolomics. In this field NMR has 2 main applications:

- direct mixture analysis of biofluids, extracts and tissues using high throughput screening essays [1-16]
- identification of individual compounds like biomarkers of disease or toxicity and drug metabolites in these mixtures by LC-NMR [17-25]

Metabonomics/Metabolomics deal with observing the entirety of molecules in biofluids, cell and plant extracts, plants and tissue samples and their compositional changes based on internal or external disturbances. Observing changes in metabolite composition is, however, only looking at the end of a chain. Such a chain starts with genetic changes that lead to changes of the proteome which induces changes in the metabolom. The whole process is known under the keyword Systeomics.

The challenge for NMR is threefold: On one side, throughput of the instruments has to be increased drastically to be competitive in the screening scenario and keep cost of analysis low; on another side, fully automatic data evaluation is needed, since with such a large amount of data, manual inspection has to be reduced to samples where the automation fails to give a clear result. Especially statistical methods are applied in the automatic analysis. The third challenge is for LC-NMR to detect compounds with much lower concentrations in a mixture, as was possible so far. Reducing the gap to UV or MS sensitivity is vital. In the sections below, the following subjects will be discussed and applications from the food area are given:
- Improvements in the hyphenation of NMR and liquid chromatography
- Combination of NMR with high throughput liquid handling
- Application of automatic statistical methods to high throughput NMR data

2 IMPROVEMENTS IN LC-NMR

Over the last 2 years LC-NMR has experienced enormous progress in versatility and mainly in NMR sensitivity. This was achieved by improvements in the interface hardware between chromatography and NMR and by NMR probehead advances.
LC-NMR systems to date could perform the following actions:

2.1 On-Flow

The output of the LC-columns is guided through the NMR flow probe and repetitive NMR acquisitions are taken with equidistant timing along the chromatography axis. For every acquisition several scans are accumulated, typically between 4 and 16 in order to not compromise on the retention time resolution. Also LC-gradient runs limit the amount of scans that can be taken for reasons of solvent suppression (NMR chemical shift depends on the solvent composition).

On-flow experiments can therefore only detect the major compounds in a mixture and columns would have to be overloaded to see more compounds.

2.2 Direct stop flow

As explained for on-flow, the time during which an LC-peak flows through the NMR detection cell is too short to allow long accumulations or even 2D-experiments. Therefore, the direct stop flow technology was developed, making use of an additional detector like UV. Under calibrated conditions it is possible to stop the flow exactly, when a LC-peak has reached the NMR flow cell and is exactly centred. This allows for running long-term accumulations and 2-dimensional experiments. The disadvantage is that if further peaks in the same chromatogram have to be investigated, the separation degrades, the more stops are performed.

2.3 Loop collection

By transferring LC-peaks intermediately into small storage capillaries, the stop flow problem can be circumvented, however retaining the LC-peak shape by using capillaries with 0.25mm i.d. in the loop collector. Versions with 12, 16 and 36 loops are currently available. Using loop storage instead of a stop flow action, only valves are switched to transfer the peaks, so the flow is never stopped during the chromatographic run and the separation performance is not degraded. Some peak broadening occurs, however, when the LC-peaks flow through the LC-NMR interface valves and additional capillaries and in the NMR flow cell.

In all operation modes mentioned so far, partly deuterated solvents are useful. To keep cost reasonable in reversed phase separations and to obtain a better observable NMR window, H_2O is normally replaced by D_2O. Methanol or Acetonitril are used in the nondeuterated form, as the deuterated solvents are too expensive. Using D_2O is a disadvantage when a mass spectrometer is added to the flow path, as exchange processes take place with hydroxyl or amino groups and typically a molecular weight distribution is obtained.

Thus far, state-of-the-art is described. In the following sections the improvements are explained.

2.4 Column switching

Having a single column for automatic separations is a problem, since different samples might need different types of columns. Further to this, when investigating biofluid compounds over the whole polarity range are eluting; however, separation is far from perfect. By introducing column switching (or 2D chromatography), both problems can be solved.

Figure 1 shows a flow chart of 2 possible valve settings. Valves in the upper chart are set so that only one column is passed by the chromatographic flow, detection occurs with the main UV cell (UV1). This corresponds to the typical 1 column situation. In the lower chart a situation is displayed where LC-peaks can be diverted from column C7 to column C1. In this case a second UV-detector that is located in the outflow of column C7 is used, while the outflow of column C1 is observed with the main UV detection cell.

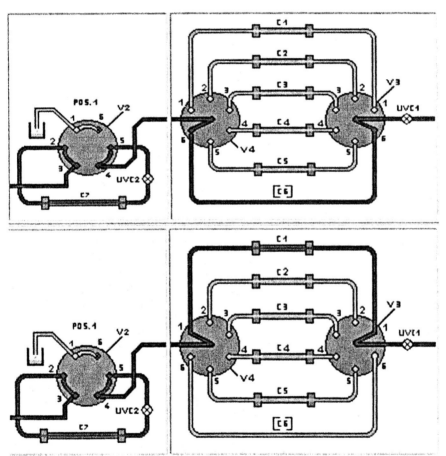

Figure 1   *Flow charts showing column switching possibilities for LC-NMR for one- or two-dimensional chromatography coupled to LC-NMR using 2 UV detection cells*

The hardware setup explained allows multiple LC-NMR applications as explained in Figure 2. In all cases the first column carries C18 reversed phase material.

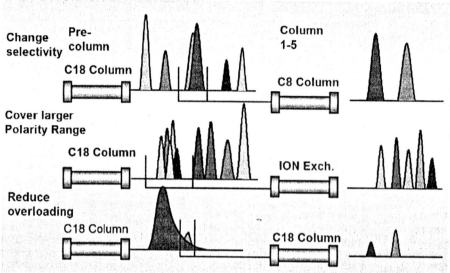

Figure 2   *Typical LC-NMR situations, where column switching is needed to improve peak separation*

The upper situation shows how to separate peaks on a second column that need different selectivity by going from C18 to C8 material. Not well separated on the first column, this co-elution can be corrected on the second column and pure LC-peaks can be transferred to the NMR.

The middle situation is typical for separations of biofluids like urine or plasma. On the reversed phase column most sugars and acids (amino and organic) elute with the dead volume of the column. If this fraction is transferred to an ion exchange column for later separation, the problem is solved. The elution from the reversed column therefore starts with pure aqueous phase. Once the polar fraction is eluted and diverted to the ion exchange column, the second column is switched out of line and the organic solvent gradient starts to elute the less polar compounds from the RP-18 column. After all peaks are eluted, the RP-18 column is switched out and the solvent system is changed to ion exchange conditions, then the ion exchange column is switched in line and the separation finished. This procedure allows a fully automatic total separation of biofluids.

The lower setup shows thus far unsolvable problem often occurring in LC-NMR: a small LC-peak that elutes in the tailing of a major peak has to be investigated. In order to obtain NMR visible concentration, often columns have to be overloaded and tailing has to be tolerated. With the column switching it is possible to let the major peak go to waste and switch a second column of the same type in line just before the small peak elutes. Once the small peak (detected in UV or MS) has eluted onto the second column it is switched out of line. After the main separation is finished, the second column is brought back in line and the separation continued on the second column, however now under the favourable condition that concentration of both peaks are much closer and no overload from the previously major peak occurs. The small peak can now be transferred to the NMR in pure form. To obtain optimal results, valves and capillaries with smaller bores are used (0.15 mm). This retains the LC-peak shape. The valves used are high pressure versions.

2.5 Loop collection improvements

As mentioned in section 2.4, smaller capillary and valve bores are essential to better NMR performance. Every LC-peak broadening has the consequence of reduced NMR sensitivity, as a smaller fraction of the peak can be placed in the active volume of the probe (volume that the detection coils observe). A further S/N improvement is achieved by drying the flow path to and inside the NMR probe prior to transfer from the sample loop. In doing so, the intermixing of the LC-peak with the preceding solvent that especially occurs when flowing into the NMR flow-cell can be reduced substantially.

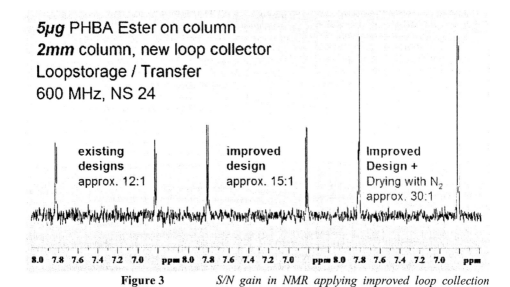

Figure 3   *S/N gain in NMR applying improved loop collection hardware*

As shown in Figure 3, NMR-S/N can be doubled with the measures explained. The spectra from left to right show the increase in sensitivity from the previous version of loop collector (left spectrum) to the new version with the smaller bores (middle spectrum) to the new version with a drying process before loop transfer (right spectrum). Spectra show the aromatic part of the spectrum of Para-hydroxybenzoicacid-n-propylester after injection of 5 microgram on a reversed phase RP-18 column (150*2mm) and an NMR acquisition at 600 MHz with 24 scans (3mm probe, 60 microliter active volume).

The valves of an improved loop collection system are high pressure models. This is needed to be compatible with post column solid phase extraction as explained in the next section. It can also be shown that cooling the loop collector allows to retain LC-peak shape for more than a week without losing NMR performance, even with solvent systems containing up to 90% CH_3CN.

2.6 Post column Solid Phase Extraction (SPE)

Solid phase extraction is normally applied prior to a separation to preclean or to concentrate compounds of interest. When integrated into LC-NMR, the SPE unit is placed behind the LC-column and the detection systems. This is equal to the loop collector position.

The SPE system brings several advantages to LC-NMR [26-28]. The first point to mention is that the separations can be carried out using standard nondeuterated solvents and buffers. This is due to the fact that the original solvents are removed from the trap cartridges by drying with nitrogen gas. Before drying, a washing step is often needed to remove salts. To enable trapping also at higher organic solvent ratio, the LC-peak flowing towards the cartridge is diluted with water.

This is also a prerequisite of multiple trapping. This means that multiple injections are done and the same peak is trapped on the same cartridge. After drying, all peak material trapped is transferred to the NMR. Multiple trapping needs an auto sampler.

A flow chart of the SPE process is shown in Figure 4. Again, as in loop collection, drying the flow path to and within the NMR before transfer is a prerequisite for highest NMR sensitivity.

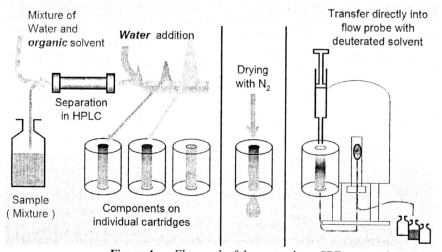

Figure 4   *Flow path of the post column SPE process for LC-NMR*

Being able to use nondeuterated solvents reduces cost and removes the deuterium exchange problem for the MS-detection. It also means that it is not necessary to reproduce the separation with deuterated solvent.

Another advantage to the SPE is the fact that the trapping step leads to an increase in S/N for the NMR that typically is in the range of a factor 2 to 4. The reason for this gain is the fact that the LC-peak is concentrated on the SPE cartridge and eluted in a small volume. The S/N gain for the NMR depends on the ratio of eluting peak volume from the LC-column compared to the eluting volume from the SPE cartridge. When using a 2mm cartridge (10mm high), the eluting volume is about 25 microliter. To achieve this low volume, mostly pure organic solvent is used for elution. As a total of about 200 microliter of solvent is needed per elution, it can be used in deuterated form and therefore improves the spectral quality in the NMR.

It is also obvious that a NMR flow cell matched in active volume to the eluting volume from the cartridge is needed to recover full sensitivity. Therefore a probe with 30 microliter active volume had to be developed. It is also obvious that with such low volume LC-peaks, capillaries and valves in the transfer pathway need to have small bore size to retain the LC-peak shape and therefore also the NMR sensitivity. A bore of 0.1mm is adequate.

The SPE approach also is more forgiving towards non-optimal chromatography, as still the whole peak can be trapped as long as there is no overlap with other peaks. In normal LC-NMR, reduction of chromatographic quality leads directly to lower sensitivity. Multiple trapping leads to a linear increase in NMR sensitivity, as long as the holding capacity of the cartridge is not exhausted. Using a second UV-cell in the outlet of the SPE cartridge allows the detection of a breakthrough.

Another advantage of the SPE system is the reproducibility of chemical shifts due to the exclusive use of organic solvents for elution. In LC-NMR, different gradient ratios lead to different chemical shift distributions of the sample signals. This is a problem for automatic matching routines that compare spectra in a database with the peaks actually measured. This is especially important in natural product investigations. The SPE system introduced is based on the Spark Prospekt II and allows a total of 192 trap cartridges to be handled during an automatic run. The system has 2 flow paths where a trap cartridge is included. This means that also 2 peaks close in retention time can be trapped without problem. To change a cartridge in the flow path typically takes about 15 seconds.

Figure 5 shows a LC-SPE-NMR/MS system at work. All relevant parts are annotated. The small pump on the upper left part is used to dilute the flow to the cartridges with water. The NMR/MS interface (called BNMI) contains the splitter that takes 5% of the flow and guides it to the MS, while 95% of the flow is directed to the NMR. The mass spectrometer is a desk top ion trap directly connected through the NMR/MS interface.

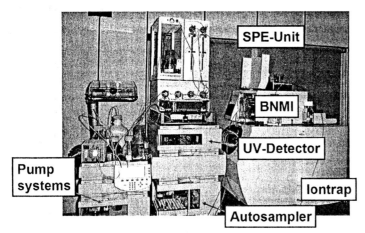

Figure 5   *Integrated   LC-SPE-NMR/MS   system   for   full automation   of   multiple   samples   and   multiple injections*

Figure 6 shows a comparison between a loop collection result (top spectrum), a single SPE trapping (middle spectrum) and a fourfold trapping (bottom spectrum) using an apple peel extract sample.

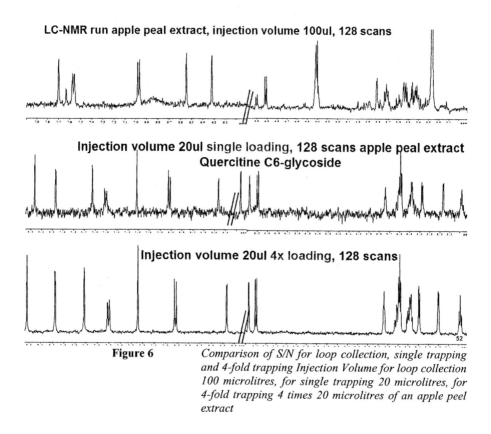

LC-NMR run apple peal extract, injection volume 100ul, 128 scans

Injection volume 20ul single loading, 128 scans apple peal extract Quercitine C6-glycoside

Injection volume 20ul 4x loading, 128 scans

Figure 6   *Comparison of S/N for loop collection, single trapping and 4-fold trapping Injection Volume for loop collection 100 microlitres, for single trapping 20 microlitres, for 4-fold trapping 4 times 20 microlitres of an apple peel extract*

For the conventional loop collection 100 microliter were injected on the column. The spectrum obtained with 128 scans shows a Quercitine Glycoside. For the single trapping only 20 microliter of the extract was injected ($1/5^{th}$ of the loop collection experiment). The spectrum obtained shows additional lines due to OH protons, which are not visible in the loop collection spectrum, as this is obtained in CH_3CN/D_2O mixture due to exchange. The spectrum after single SPE is obtained in CD_3CN, where no exchange happens. Another advantage of the SPE system is that exchangeable protons are visible and can be used for nuclear Overhauser experiments and other homonuclear 2D-experiments useful for structure determination and stereochemistry.

When exactly comparing S/N also to the fourfold trapping, an S/N increase of a factor 4 is detected as expected from single to 4-fold trapping. Going from loop collection to single trapping increases S/N by a factor of 3.8 in the NMR. It can be seen that multiple trapping can easily generate an S/N increase of more than one order of magnitude.

With such tools in hand, it is possible to run 2-dimensional experiments that are impossible with conventional LC-NMR (loop collection and direct stop-flow). Figure 7 shows the result of an overnight measurement session on Phyllanthin, a compound that was trapped after LC-separation of an extract of Phyllanthius Myrtifolius. Ten microliter of a 222 microgram/microliter extract was injected onto a reversed phase column. Having the 2D spectra together with the mass spectrum in many cases allows for the determination of structures immediately.

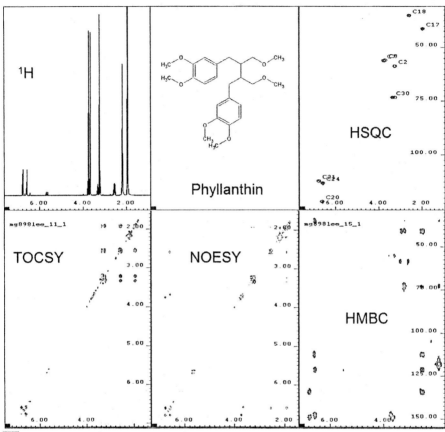

Figure 7 *2D-spectra obtained from a LC-SPE-NMR/MS run in an overnight measurement on Phyllanthin, isolated from Phyllanthius Myrtifolius in a single SPE trapping step*

2.7 Cryogenic NMR probes [26, 29-39)]

NMR sensitivity can be raised by using high magnetic field strength; however, the cost of a superconducting magnet increases exponentially when going to fields above 700 MHz. Therefore, other means to increase S/N are investigated. Sensitivity is also defined by the NMR probehead and the signal amplification used.

By reducing the temperature of the detection coil, the NMR sensitivity can be raised. This is the principle of the cryogenic probes. While the sample stays at ambient or desired temperature, the detection coils are operating at a temperature of about 20K, thus reducing the thermal noise and increasing NMR sensitivity by an average factor of four. Besides the coils, also the preamplifiers are cooled to ~ 70K. The cooling is achieved by cycling cold gaseous helium in a closed system.

While the first cryogenic probes introduced were tube-based, in 2002 dedicated flow cryoprobes were made available. The technology has further evolved and probes available today have flow inserts that allow converting from NMR tubes to flow applications in a few minutes.

When using LC-NMR there is a demand for a small volume detection cell, which is ideally combined with a so-called micro-cryoprobe allowing 3mm tubes as the maximum size. Flow cells for 10, 30 and 60 microliter of active volume are available for this type of probe. The 30 microliter version is a special match to the eluting volume from a SPE cartridge of 2mm inner diameter. Ten microliter cells are tailor made for SPE cartridges of 1mm inner diameter. Having the possibility to exchange the flow insert increases the flexibility of such expensive probes and allows to adjust to the special measurement problem. A 60 microliter cell is a good compromise for on-flow/stop-flow work with 2mm LC-columns and the SPE. The sensitivity gain combining SPE and cryogenic probe with flow insert can be seen in Figure 8.

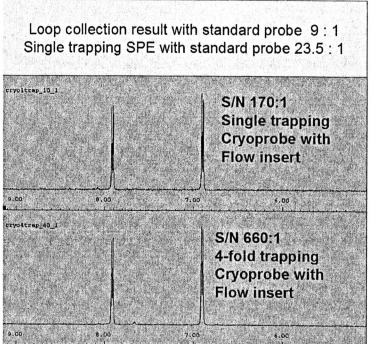

Figure 8   *Comparison of 500 MHz S/N values comparing loop collection, single trapping and fourfold trapping with room temperature and cryogenic probe with flow insert on an injection of 5 micrograms of Para-hydroxybenzoicacid-n-propylester onto a reverse phase column*

An injection of 5 microliter of Para-hydroxybenzoicacid-n-propylester on column (RP-18, 150*2mm) is analysed first with loop collection (16 scans). An S/N of 9:1 is achieved on the aromatic signals using a 60 microliter standard probe with room temperature detection. On the same probe, SPE single trapping leads to an S/N of 23.5:1. When the cryogenic probe with 30 microliter flow insert is used for the single trapping, the S/N goes up to 170:1. By doing 4-fold trapping the S/N with the cryoprobe system is brought up to 660:1. This numbers show the enormous potential of the combined technology. With multiple trapping up to 2 orders of magnitude increase in S/N are possible and enable NMR to detect low concentration compounds that could not be analysed before. 2D-experiments are accessible within a short time.

Carvacrol is identified in an Acetone Oregano extract using the LC-SPE-NMR/MS setup with the cryoprobe. Only the cryoprobe also allows recording a carbon direct detected spectrum after the SPE trapping. In such a case, inverse experiments can be acquired even with single scans per increment. The spectra obtained on Carvacrol during a reversed phase separation are shown in Figure 9. The peak again is detected by mass spectroscopy in the chromatogram and transferred to the SPE cartridge. A triple trapping is applied.

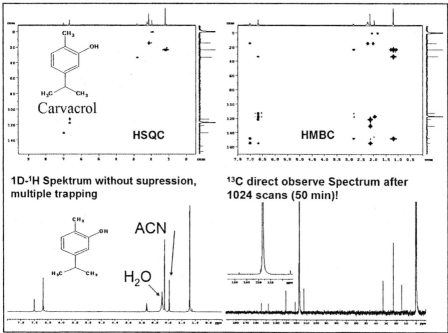

Figure 9   *Carvacrol spectra obtained from an Acetone Oregano extract after triple trapping and transfer to the cryogenic probe with flow insert. The carbon spectrum was obtained with 1024 scans at 600 MHz*

In molecules with few protons, long range inverse correlations might not allow to determine the structure; therefore, direct carbon observation is a possible solution, which can be applied to LC-NMR samples as well. It should be noted that the results shown are obtained on an inverse probe, where the inner coil is tuned to proton and the outer coil to carbon. Carbon spectra can be a factor of 2 more sensitive if a carbon-observe probe is used. In this case, the proton sensitivity has to be compromised by a factor of 2 compared to the inverse configuration.

3 HYPHENATION OF NMR WITH HIGH-THROUGHPUT FLOW INJECTION[6,40]

LC-NMR can be considered as a low-throughput application, because of the longer acquisition times needed due to the limited amount of sample that can be injected on to a chromatographic column. When a liquid sample, however, is directly injected into the flow cell, turnover can be very high. This is the typical situation in biofluid screening. Besides the speed advantage, there is no need to refill a sample into a NMR tube. Due to the height of standard NMR tubes, they cannot be filled by a standard liquid handler. It needs very expensive hardware to solve the problem.

In flow injection NMR, vials or well plates can be used directly. In addition, such liquid handlers can also do the sample preparation. In biofluid applications it is necessary to add buffers to adjust the pH of the sample to a fixed value throughout a study.

Chemical shift depends strongly on pH for most small molecules under investigation. Therefore, a typical scenario is to take out an aliquot of the sample into an empty vial, add buffer to it and then mix it efficiently. This preparation can be done offline or inline with the measurement. If offline preparation is done, it is important to cool the prepared samples until they are transferred to the NMR. Special care has to be taken with well plates to avoid a temperature gradient in the well. Special peltier elements with cooling fingers that stick into the well plate from the bottom side have been developed. Using 96-well plates, a total of 960 samples can currently be handled on the bed of a standard liquid handler as used in the flow injection NMR setup.

Measurement of samples is typically executed in the range of 298 to 313 K. The cooled sample has a temperature of 277 K. If the sample is transferred cold into the flow cell of the NMR probehead, it would take several minutes to adjust a homogeneous temperature throughout the cell. This is a waste of time and against the high-throughput screening idea. Therefore, a heated transfer line is used that brings the sample to measurement temperature during the transfer. The overshoot needed in this process can be reduced by also heating the transfer line inside the probehead from the input connector to the flow cell. When the temperature is calibrated, measurement in the NMR flow cell can be started as soon as the transfer is finished. To allow high transfer speed, capillaries with 0.5mm inner diameter are used to reduce back pressure. Typical transfer speeds can range up to 15ml/min.

A further advantage of flow injection NMR is the substantially reduced need for shimming the magnet between samples. This is due to the fact that the flow cell is a fixed design, not like tubes, that are exchanged for every sample and always show some slight irregularities. Turnover can be ranging from less than a minute to 5 or 6 minutes for experiments needing more scans. The buffer can be prepared in D_2O if 2D-measurements are also planned, for short measurements, deuterated solvents can be omitted.

A special integrated spectrometer for screening processes is shown in Figure 10. The magnet is situated inside the enclosure, as is the NMR console. The liquid handler is placed on the working bench. The magnet in this case is a double shielded design that, for example, at 400 MHz guarantees that the 5 gauss line is inside the magnet dewar. Thus the disturbance through the magnet is reduced to a minimum; even persons with pacemakers can be in the direct vicinity of such an instrument. This concept of extra shielding allows to place the liquid handler directly next to the magnet and to minimize transfer pathways and thereby also time needed to transfer samples from the liquid handler to the flow cell.

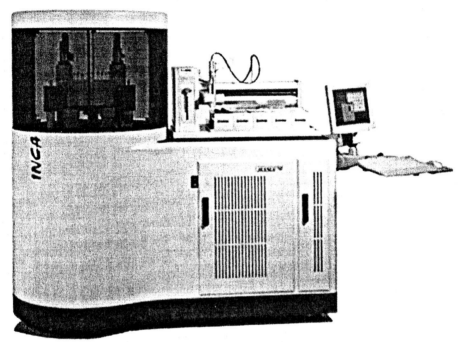

Figure 10   *400 MHz screening machine for flow injection NMR with special ultra-shielded magnet*

Back mixing between samples is minimized by cleaning segments and drying of the transfer pathway. Sample segments can be packed in between gas and liquid gaps to separate from transfer and cleaning solvent. Back mixing is typically brought below 0.5%. Further reduction is possible by extra cleaning efforts; however is time consuming. Flow Injection NMR is always a compromise between performance and speed.

Typical applications for flow injection NMR are found in the area of toxicity screening of new drug candidates, coronary heart disease risk assessment, natural product screening and combinatorial chemistry/parallel synthesis. Other possible application areas are in food quality control and clinical disease screening.

4 NMR COMBINED WITH STATISTICAL METHODS

Previously, NMR was considered as a tool for structure elucidation of pure compounds, but recently the investigation of mixtures has become an important application. This is due to the high-speed screening capability of NMR flow injection systems and the existence of software tools to investigate mixture spectra automatically.

While statistics has been a standard tool to investigate NIR spectra for many years, NMR just came into the game in recent years and was driven by the rapidly growing research field of Metabonomics. Most of the applications use statistics on 1D-proton spectra, which can be rapidly obtained. While the investigation of urine and plasma is already established in the pharmaceutical environment, examples from the area of food quality control are given here.

One question of importance is the ability to control mixing of different fruit juices. It is, for example, possible to add up to 15% pear juice to apple juice before this can be tasted. In multi-fruit juices it is necessary to be able to check the mixing ratios given on the label of a product. A more challenging problem is the differentiation of direct orange juice from rediluted concentrate. Figure 11 shows a PCA analysis [41-43] of direct juices from Brazil, Mexico, Morocco and Spain, pasteurized juices and rediluted concentrates of the same juices.

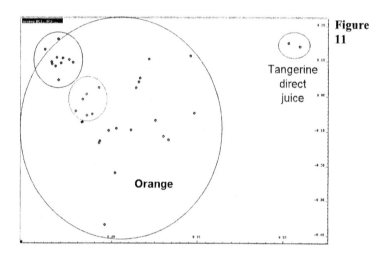

Figure 11   *PC1 versus PC2 map of a set of 1D-proton 400 MHz NMR spectra of orange juices obtained from Brazil, Mexico, Morocco and Spain. Direct juices and their corresponding rediluted concentrates are included as well as some pasteurized juices*

The PC1/PC2 map is obtained from flow injection 1D-proton spectra at 400 MHz, where the water signal is suppressed by the 1D-version of the 2D-NOESY sequence. All spectra included in the analysis were prepared under identical conditions. The input into the statistical calculations is done by bucketing of the spectra; this means to subdivide the spectrum into equidistant chemical shift segments, calculate the integral for each segment and to do auto scaling. Shift regions of variation not related to the question asked are excluded from the calculation; in this case, the region around the suppressed water is taken out of the calculation. By varying the shift region used for calculation it can be shown that there are two spectral regions, one in the aromatic part of the spectra and one in the region between 3.5 and 2.5 ppm that especially explain the separation of the groups. The differences to be observed can be traced back to the thermal treatment needed to obtain the pasteurized juices and the concentrate.

So far, no other analytical method could identify the differences due to treatment. It is obvious that there is room for fraud, since there is, for example, a factor of ~ 5 in volumes whether a direct juice or a concentrate is transported. Therefore, the direct juice is higher in cost and money could be saved by mislabelling or blending.

A second example is shown in Figure 12. Here the task aimed at is differentiation of top and bottom fermented beers. As can be concluded from the PCA map, separation occurs on the type of malt used rather than on the type of fermentation. This explains why the Alt, which is a top fermented beer, actually falls into the bottom fermented beers region (Pilsner, Stark/Bock). Alt is made from barley malt, as is true for Pilsner and Stark/Bock-beer. Wheat-beer is made from wheat malt. Spectra used are obtained at 400 MHz with flow injection NMR and triple suppression of water and ethanol signals. Only the aromatic region was used for bucketing.

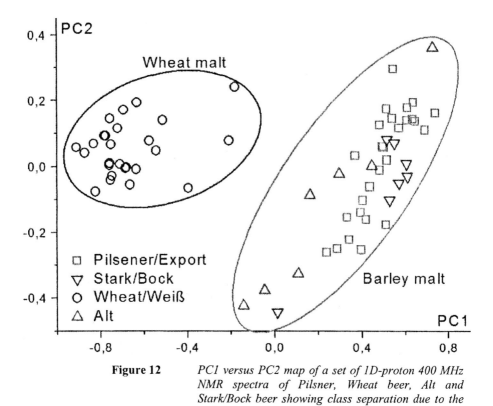

Figure 12   *PC1 versus PC2 map of a set of 1D-proton 400 MHz NMR spectra of Pilsner, Wheat beer, Alt and Stark/Bock beer showing class separation due to the malt type used*

This test can be used to control the purity of the beer, as, for example, defined by law in Germany.

Besides the results based on principal component analysis, PLS [42,43] can also be used. The typical case is mentioned in the beginning of this chapter with the recognition of juice mixing. PLS can further be used to determine the original gravity in beer.

In German beer legislation, beer categories depending on the content of original gravity are defined. Original gravity is the concentration of solids in the unfermented wort from which the beer is made. Traditionally, it is calculated from real extract and alcohol content of the beer. In the NMR spectrum single substances cannot be chosen as characteristic for original gravity because the malt extract contains a complex range of ingredients. For the determination of original gravity, the bucket table of the entire NMR range (10.0-0.3 ppm) including ethanol can be correlated to the wet chemical analysis using PLS. The PLS model, verified through cross-validation, shows an excellent correlation (R=0.998, p<0.0001) between the standard laboratory values and the NMR prediction.

Another parameter which must be quantified in the context of the official food monitoring is the ethanol concentration. By directive in the European Union, maximum tolerances of the indication of the alcoholic strength in the labelling are specified. The PLS model for ethanol (1.29-1.05 ppm) also shows a good correlation (R=0.985, p<0.0001). None of the beers is false-positively out of the tolerance. The NMR prediction is therefore able to efficiently control the legal tolerances of ethanol.

It is obvious that in the same way as ethanol is quantified using PLS, also other ingredients like amino acid or organic acids can be quantified out of the same experiment. Therefore, it is important to mention that the NMR measurement can replace a whole suite of wet tests and by this reduce the cost of the analysis.

5. CONCLUSIONS

It has been shown that NMR has expanded into the field of mixture analysis. Two main strategies are followed:
-    isolation and identification of individual compounds in a mixture by LC-NMR
-    direct mixture analysis using flow injection high-throughput NMR
The LC-NMR approach could be improved substantially in performance by the introduction of optimized hardware, post column solid phase extraction and cryogenic probes with flow insert. Taking all tools together, S/N can be improved by up to 2 orders of magnitude compared to the status about 2 years ago.

This increase in S/N enables the investigation of very low concentrated LC-peaks and still be able to run 2-dimensional experiments as needed for structure elucidation. Structure elucidation is further helped by the integrated NMR and MS analysis. With the cryoprobe technology it is also possible to obtain carbon spectra directly from LC-NMR experiments. Flow-Injection NMR has enabled high-throughput screening and as a consequence the efficient use of statistical analysis on NMR data. It could be demonstrated that the NMR analysis can generate not only statistical data, but also deliver quantitative results on a variety of small molecules that are used, for example, to characterize quality.

Acknowledgments

P. Rinke at SGF in Germany has donated all orange juice samples and helped with data evaluation.
D.W. Lachenmeier from the CVUA in Karlsruhe, Germany has donated the beer samples investigated and helped with data evaluation.
A. Lommen from Rikilt DLO in Holland has provided the apple peel extract sample.
The Oregano extract was obtained from J. Vervoort from the Wageningen NMR centre in Holland.
The Phyllanthius Myrtifolius extract was obtained from S.Lee, NTU, Taipei, Taiwan.

References

1. J.C.Lindon, E.Holmes and J.K.Nicholson, Prog. NMR Spectrosc. **45** (2004) 109
2. J.K.Nicholson, J.Connelly, J.C.Lindon, E.Holmes,
   Nat.Rev. Drug Discovery (2002) Vol.1, 153
3. R.Soper, U.Himmelreich, D.Painter, R.L.Somorjai, C.L.Lean, B.Dolenko,
   C.E.Mountford, P.Russell, Pathology, 34 (2002) 417
4. C.L.Gavaghan, E.Holmes, E.Lenz, I.D.Wilson, J.K.Nicholson FEBS Letters,
   484 (2000) 169
5. C.Zuppi, I.Messana, F.Forni, C.Rossi, L.Pennacchietti, F.Ferrari, B.Giardina,
   Clin.Chim.Acta, 265 (1997) 85
6. M.Spraul, M.Hofmann, M.Ackermann, A.W.Nicholls, S.J.P.Damment,
   J.N.Haselden, J.P.Shockcor, J.K.Nicholson, J.C.Lindon,
   Anal.Comm. Vol.34, (1997), 339
7. M.J.Lynch, J.Masters, J.P.Pryor, J.C.Lindon, M.Spraul, P.J.D.Foxall, J.K.Nicholson
   J.Pharm.Biomed.Anal. Vol.12, Nr.1 (1994), 5
8. I.F.Duarte, A.Barros, C.Almeida, M.Spraul, A.M.Gil, J.Agric.Food.Chem.
   2004, 52, 1031
9. I.Duarte, A.Barros, P.S.Belton, R.Righelato, M.Spraul, E.Humpfer, A.M.Gil,
   J.Agric.Food.Chem. **2002**, 50, 2475
10. L.Mannina, M.Patumi, N.Proietti, D.Bassi, A.L.Segre, J.Agric.Food.Chem.
    2001, 49, 2687
11. I.J.Colquhoun, Spectr.Europe 10/1 (1998), 8
12. P.S.Belton, I.J.Colquhoun, E.K.Kemsley, I.Delgadillo, P.Roma, M.J.Dennis,
    M.Sharman, E.Holmes, J.K.Nicholson, M.Spraul, Food Chem. Vol.61 (1998), 207
13. B.Sitter, U.Sonnewald, M.Spraul, H.E.Fjösne, I.Gribbestad,
    NMR Biomed 2002, **15**, 327
14. S.Garrod, E.Humpfer, M.Spraul, S.C.Connor, S.Polley, J.Connelly, J.C.Lindon,
    J.K.Nicholson, E.Holmes, Mag.Res.Med., Vol.41, (1999), 1108
15. D.Moka, R.Vorreuther, H.Schicha, M.Spraul, E.Humpfer, M.Lipinski, P.J.D.Foxall,
    J.K.Nicholson, J.C.Lindon, Anal.Comm., Vol.34 (1997), 107
16. A.M.Gil, I.F.Duarte, I.Delgadillo, I.J.Colquhoun, F.Casuscelli, E.Humpfer,
    M.Spraul, J.Agric.Food Chem. **2000**, 48, 1524
17. O.Corcoran, M.Spraul, Drug Discovery Today, Vol.8 (2003) 624
18. G.B.Scarfe, B.Wright, E.Clayton, S.Taylor, I.D.Wilson, J.C.Lindon, J.K.Nicholson,
    Xenobiotica, **1999**, 29, 77
19. E.Clayton, S.Taylor, B.Wright, I.D.Wilson, Chromatographia, **1998**, 47, 264
20. O.Corcoran, M.Spraul, M.Hofmann, I.M.Ismail, J.C.Lindon, J.K.Nicholson,
    J. Pharm. Biomed. Anal., 16 (1997), 481
21. J.P.Shockcor, S.E.Unger, I.D.Wilson, P.J.D.Foxall, J.K.Nicholson, J.C.Lindon
    Anal.Chem., **1996**, 68, 4431
22. A.E.Mutlib, J.T.Strupczewski, S.M.Chesson, Drug.Metab.Dispos. **1995**, 23, 951
23. M.Spraul, M.Hofmann, J.C:Lindon, D.Farrant, M.J.Seddon, J.K.Nicholson,
    I.D.Wilson, NMR Biomed., **1994**, 7, 295
24. M.Spraul, M.Hofmann, P.Dvortsak, J.K.Nicholson, I.D.Wilson,
    Anal.Chem., **1993**, 65, 327
25. M.Spraul, M.Hofmann, I.D.Wilson, E.Lenz, J.K.Nicholson, J.C.Lindon,
    J. Pharm. Biomed. Anal. **1993**, 11, 1009
26. V.Exarchou, M.Godejohann, T.V.Beek, I.P.Gerothanassis, J.Vervoort,
    Anal.Chem., **2003**, 75, 6288

27. O.Corcoran, P.S.Wilkinson, M.Godejohann, U.Braumann, M.Hofmann, M.Spraul, American Laboratory: Chromatography Perspectives, **Vol.5**, 2002, 18
28. L.Griffiths, R.Horton, Mag.Res.Chem., Vol.36 **(1998)**, 104
29. A.Eletsky, O.Moreira. H.Kovacs, K. Pervushin, J. Biomol. NMR, 26 **(2003)** 167.
30. M.Spraul, A.S.Freund, R.E.Nast, R.S.Withers, W.E.Maas, O.Corcoran Anal.Chem., **2003**, 75, 1536
31. J.L.Griffin, H.C.Keun, C.Richter, D.Moskau, C.Rae, J.K.Nicholson, Neurochemistry International, January **2003**, vol. 42, iss. 1, pp. 93-99(7)
32. H.C.Keun, O.Beckonert, J.L.Griffin, C.Richter, D.Moskau, J.C.Lindon, J.K.Nicholson Anal Chem 74(17), 4588-93 **(2002)**
33. C.A.Lepre, J.Peng, J.Fejzo, N.Abdul-Manan, J.Pocas, M.Jacobs, X.Xie, J.M.Moore, Comb.Chem.High Throughput Screen., 5 **(2002)** 583.
34. J.Boisbouvier and A.Bax, J.Am.Chem.Soc., 124 **(2002)** 11038.
35. D.Monleón, K.Colson, H.N.B.Moseley, C.Anklin, R.Oswald, T.Szyperski, G.T.Montelione, J.Struct.Funct.Genetics, 2 **(2002)** 93.
36. P.J.Hajduk, D.J.Augeri, J.Mack, R.Mendoza, J.Yang, S.F.Betz S.W.Fesik, J.Am.Chem.Soc., 122 **(2000)** 7898.
37. A.Medek, E.T.Olejniczak, R.P.Meadows, S.W.Fesik, J.Biomol.NMR, 18 **(2000)** 229
38. P.J.Hajduk, T.Gerfin, J.-M.Boehlen, M.Häberli, D.Marek, S.W.Fesik, J.Med.Chem., 42 **(1999)** 2315.
39. P.Styles, N.F.Soffe, C.A.Scott, D.A.Cragg, D.J.White and P.C.J.White, J.Magn.Reson., 60 **(1984)** 397.
40. M.Spraul, E.Humpfer, S.Keller, H.Schäfer, Bruker BioSpin Spin Report 154/155, **2004**, 26
41. J. Edward Jackson, A User's Guide to Principal Components John Wiley & Sons, Inc. New York Chichester Brisbane Toronto Singapore, 1991
42. D.L.Massart, B.G.M.Vandeginste, L.M.C. Buydens, S. De Jong, P.J. Lewi, J. Smeyers-Verbeke, Handbook of Chemometrics and Qualimetrics: Part A Elsevier Amsterdam Lausanne New York Oxford Shannon Singapore Tokyo 1997
43. B.G.M. Vandeginste, D.L. Massart, L.M.C. Buydens, S. De Jong, P.J. Lewi, J. Smeyers-Verbeke, Handbook of Chemometrics and Qualimetrics: Part B Elsevier Amsterdam Lausanne New York Oxford Shannon Singapore Tokyo 1997

MAGNETIC RESONANCE MEASUREMENTS OF STRUCTURAL CHANGES DURING HEATING OF CHICKEN MEAT BY HOT AIR

S.M. Shaarani, K.P. Nott and L.D. Hall

Herchel Smith Laboratory for Medicinal Chemistry, University of Cambridge School of Clinical Medicine, Robinson Way, Cambridge, CB2 2PZ, United Kingdom

1 INTRODUCTION

The water content of meat is of great importance because it is the major component (65-80%); furthermore, its interactions with macromolecules determines the meat's water holding capacity[1,2] and influences meat characteristics such as structure, stability and microbiological safety.[3] Many studies have used low-field Nuclear Magnetic Resonance (NMR) to investigate water in meat, in relation to its water holding capacity.[4,5,6] Magnetic Resonance Imaging (MRI) has been used to visualise water distribution in porcine muscles after tumbling,[7] drying,[8] and freeze-thawing.[9] However, very few studies have combined both bulk NMR with MRI to quantitate changes of structure and water in meat processing.[9]

Clearly it is preferable to measure serially and on-line the changes in temperature, water content and structure that occur during cooking. Hence, the purpose of this study was to explore the potential for using that combination of NMR and MRI to follow the heating of chicken meat by hot air. Specifically, the above involved:

a) comparison of the values of water content (M_o) and its spin-spin (T_2 values) and spin-lattice (T_1 values) relaxation times obtained by NMR and MRI
b) establishing the relationship between gravimetric moisture content and various MR parameters
c) using those data to understand the series of MR images obtained during heating by hot air.

2 METHODS

2.1 Off-line NMR and MRI experiments

2.1.1 Sample Preparation and Cooking Procedure. Chicken meat was cut into $4 \times 4 \times 1$ cm slabs (each approximately 15.5 gram). A convection oven (Sharp, model R-84STM) was first pre-heated to 200°C and then seven samples were heated on the 'grill-shelf' with turn table rotation for 3, 6, 9, 12, 15, 18 and 21 minutes respectively and then left to cool in a dessicator at room temperature. Samples were weighed before and after cooking at room temperature.

2.1.2 Moisture Content Determination. Moisture content was measured by gravimetric loss of water from the sample that had been dried by heating in an oven at 103°C and left over night.[10]

2.1.3 MRI Hardware and Quantitation. All MRI measurements were acquired using a 2.35 Tesla, 31cm horizontal-bore superconducting magnet (Oxford Instruments, Oxford, U.K.) connected to a Bruker Medzintechnik Biospec II imaging console (Karlsruhe, Germany). Each axis of the gradient set (11.6 cm internal diameter) was powered by a pair of Techron gradient amplifiers (Model 7790, Crown International Inc., Elkart, IN, U.S.A.). A cylindrical, eight-strut, bird-cage radiofrequency (RF) probe (internal diameter of 5.9 cm), was used in quadrature mode to transmit and receive the MR signals.

Bulk NMR T_2 measurements were acquired using a Carr Purcell Meiboom Gill (CPMG) sequence with 256 points, TE 1.0 ms and TR 10 seconds. The fitting analyses were performed using Gnuplot software (Linux version 3.7).

MRI quantitation of the relaxation times (T_1, T_2) and liquid proton density (M_0) were acquired for 2D slice images (0.55 mm spatial resolution 5 mm slice thickness) and 2 averages. Sets of T_1-weighted images were acquired with an echo time (TE) 6 ms and repetition times (TR) 0.4, 0.75, 1.0, 2.5 and 5.0 s; least squares fitting of each pixel to a mono-exponential saturation recovery model gave T_1 map. Similarly a set of 16 T_2-weighted spin echo images were acquired with TE 6 ms and TR 5.0 s; least squares fitting of each pixel to a mono-exponential transverse relaxation decay model gave the T_2 maps; M_0 was obtained by extrapolating the fitted transverse decay back to zero time for each pixel. All MRI data was fitted using software written by Dr. P.J. Watson and the images were visualised using image display software (Cmrview) written by Dr N.J. Herrod.

3 RESULTS AND DISCUSSION

3.1 'Off-line' measurements

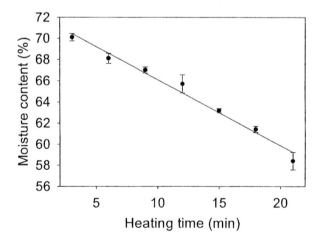

Figure 1 *Gravimetric moisture content determination of chicken meat during cooking in a convection oven at 200°C. Measured off-line at 20°C.*

The determined gravimetric moisture content for fresh chicken meat was 75%. After the meat was cooked for 3 minutes, the moisture content had reduced to 70% (Figure 1); it was further reduced to 58% after heating for 21 minutes at which time the surface of cooked meat has become dry and hard. There is a good linear correlation between moisture content with heating time (Figure 1, $R^2=0.9804$).

For the raw meat the bulk NMR T_2 relaxation decay gave a multi-exponential fitting with three distinct water populations which is in agreement with previous studies[5,11]; a long component $T_{21}=72$-104 ms (9%), an intermediate, major component $T_{22}=40$-44ms (87%) and a short component $T_{23}=2$-3 ms (4%) (Figure 2(a), (b) and (c)). For the cooked meats, the T_{21} values ranged from 62 to 139 ms, T_{22} of the major component decreased from 31 to 19 ms, and T_{23} increased from 3 to 7 ms. It has been suggested that the major component (T_{22}) is from water which is associated within the highly organised protein structures with high myofibrillar protein densities, which includes actin and myosin. The long component (T_{21}), represents water between fibre bundles as well as the inter myofibrillar water and the short component (T_{23}), has been assigned to water which is associated with macromolecules.[5,12] In general, the T_{21} values were observed to increase with heating time whereas the T_{22} values decrease. Micklander et al.[12] suggested this was due to the myosin protein denaturing, causing the myofibrils to contract, thereby expelling water from within the myofibrils (represented by T_{22}) to the inter myofibrilar spaces (represented by T_{21}) and thus a reduction in myofibrillar spacing.

The T_1 values from the imaging data of the raw meat ranged from 0.65 to 1.06 s, whereas T_2 values ranged from 31 to 45 ms; the mean values of T_1 and T_2 during cooking are plotted, along with their values prior cooking, in Figure 2(d) and 2(e), respectively. The fact that both T_1 and T_2 values decreased on heating is due to protein denaturation which causes loss of water holding capacity and hence moisture loss. The T_1 values of meat that has been heated for 3 minutes ranged from 0.74 to 1.0 s. After the meat had been heated for 21 minutes their T_1 values ranged between 0.33 to 0.49 s. The T_2 values from the imaging data of raw and cooked chicken meat after 3, 9 and 21 minutes is illustrated in Figure 2(f); their distribution is widest (ca. 13-48 ms) for the raw, then narrows (ca. 9-34 ms) by the midpoint (9 min), and finally the narrowest (ca. 11-27 ms) at the endpoint (21 min). The broader distributions of T_2 indicate that the meat is still raw/undercooked whereas the narrower T_2 distributions suggest that the meat is fully cooked.

It is interesting to note the high correlation for the cooked meat between the values obtained from the T_2 images (Figure 2(d)) and the intermediate component, T_{22}, from bulk NMR (Figure 2(b)) (r=0.982, p<0.01). This suggests that the T_2 mapping protocol used with TE=6 ms and 16 echoes effectively captured the majority, intermediate component (T_{22}) of the chicken meat measured by bulk NMR with TE=1ms and 256 echoes. The other components were either too short, or too long, to have an influence; this gives confidence that imaging values are accurate.

Table 1 demonstrates that there is a positive significant correlation (r=0.861,p<0.01) between gravimetric moisture content and T_{22}, but a negative significant correlation (r=-0.725, p<0.05) with T_{21} of cooked meat. This again illustrates the point that as the proteins denature, water is expelled from the T_{22} domain (which decreases T_{22} values) into the T_{21} domain (which increases T_{21} values). There is also a significant correlation (r=0.870, p<0.01) between the T_2 values from imaging measurements and moisture content. This is not surprising since T_2 is highly correlated with T_{22} (r=0.997, p<0.01).

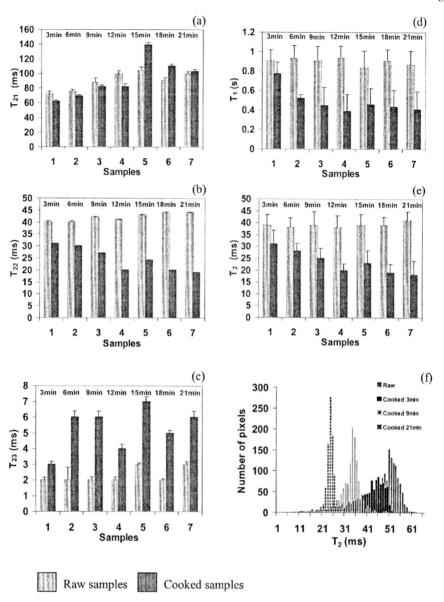

Figure 2 *Three components of bulk T_2 relaxation time (a) T_{21} (b) T_{22} (c) T_{23} and mean value of (d) T_1 and (e) T_2 from imaging (f) illustrates the range of T_2 values from the imaging for raw and cooked for 3, 9 and 21 minutes measured off-line at room temperature.*

Table 1 *Pearson's correlation analysis (single tailed) between gravimetric moisture content and various MR parameters*

Correlation Parameters	Correlation coefficient (r)	Significance level
A. Bulk NMR		
Total M_0	0.936	0.01
T_{21}	-0.725	0.05
T_{22}	0.861	0.01
T_{23}	-0.462	Not significant
B. MRI		
M_0	0.729	0.05
T_2	0.879	0.01
T_1	0.673	0.05

The high significant correlation (r=0.936, p<0.01) between moisture content and total M_0 from bulk T_2 measurement of cooked meat, strongly suggests that changes in moisture content are observed in total M_0 obtained from bulk measurement. There is a smaller correlation between moisture content and M_0 from imaging, r= 0.714 (p<0.05) probably reflecting the fact that this represents only one, albeit the major water component in meat.

3.2 'On-line' heating

Although the use of MRI to quantitate the progression of cooking on-line is beyond the scope of this study, it is appropriate to demonstrate its substantial potential by qualitative observations on uncalibrated MRI 'pictures'.

'On-line' cooking was measured in a purpose designed probe housed in a glass vessel, with a variable temperature probe (Figure 3) heated by hot air generated by flowing compressed air over an electrical heating element. The samples of chicken meat were placed on a 'sample support' at the bottom of the probe and heated at 102°C for 36 minutes.

A heating front was observed as a dark region moving from the outside to the core of the meat (Figure 4); subsequently, the intensity of the entire chicken piece decreased with heating time. It was also observed in these 2D images that the meat shrunk in one dimension while it expanded on the other. This reflects both shrinkage of the sarcomeres (units of myofibrils)[13,14] and loss of water[15] due to protein denaturation.

In the absence of other data, it is reasonable to suggest that the intensity of the chicken meat was initially predominately temperature related, whereas in the later stages it was a function of moisture content.

4 CONCLUSIONS

This study has demonstrated that both bulk NMR and MRI measurements of the T_2 and M_0 values of water reflect the structure and moisture content during heating of chicken meat. In accord with the literature, multi-exponential analysis of the bulk NMR spin-spin relaxation behaviour identified three distinct water populations. That the values of the major component (T_{22}) correlate with those taken from T_2 maps suggests that in future studies MRI can be used to measure the spatial distribution of T_2 values.

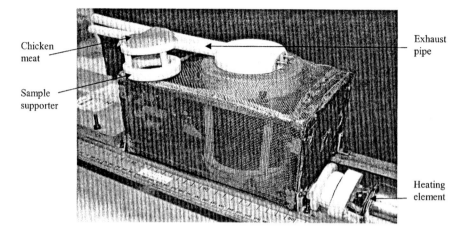

Chicken meat

Sample supporter

Exhaust pipe

Heating element

Figure 3 *The variable temperature probe with heating element*

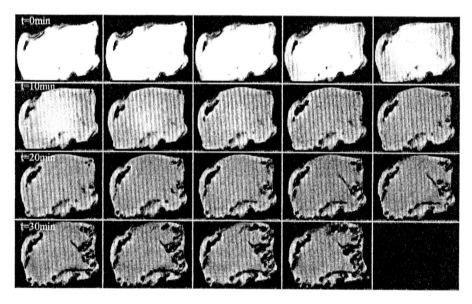

Figure 4 *A series of 2D images with spatial resolution of 0.468 μm were acquired on a 2 Tesla magnet; a gradient echo sequence with TE=7.5ms, TR= 200ms, θ= 60° and 2 averages, the total scan time 52 seconds and acquired every 2 minutes. Heated at 102°C for 36 minutes with flow of air at a rate of ca. 200 litre/minute.*

The significant correlations between gravimetric moisture measurements and various MR parameters can be used to describe moisture content in chicken meat at room temperature.

Such measurements made on-line clearly have many advantages, and future reports will demonstrate how they can not only help our understanding of the effects of heating of many foods, but also provide quantitative data to validate modelling of heating changes and associated process modelling.

Acknowledgements

LDH gratefully acknowledges the late Dr Herchel Smith for his endowment. SMS thanks the University Malaysia Sabah for his scholarship. KPN would like to thank the UK Biotechnology and Biological Sciences Research Council (BBRSC) for his grant (No. 8/D15646). We also like to thank Dr. P. Jezzard and Dr C.J. Wiggins for their legacy of hardware used for hot air heating; Richard Smith, Simon Smith and Cyril Harbird for supply and maintenance of the MRI hardware; Dr Da Xing and Dr Nicholas Herrod for the computer facilities and software respectively; and Dr Paul Watson for the curve fitting software.

References

1 J.P. Renou, G. Bielicki, J.M. Bonny, J.P. Donnat and L. Foucat, Assessment Of Meat Quality By NMR, in *Magnetic Resonance in Food Science*, eds., P.S. Belton, A.M. Gil, G.A. Webb and D. Rutledge, Royal Society of Chemistry, Cambridge, 2003, pp.161-171.

2 J.P. Renou, L. Foucat and J.M. Bonny, *Food Chemistry*,2003,**82**,35

3 S. Fjelkner-Modig and E. Tornberg, *Meat Science*, 1986,**17**,231.

4 J. Brøndum, L. Munck, P. Henckel, A. Karlsson, E. Tornberg, S.B. Engelsen, *Meat Science*, 2000, **55**, 177.

5 H.C. Bertram , A.H. Karlsson, M. Rsmussen, O.D. Pedersen, S. Dønstrup and H.J. Andersen, *J. Agric. Food Chem.*, 2001, **49**, 3092.

6 H.C. Bertram, S. Dønstrup, A.H. Karlsson and H.J. Andersen, *Meat Science*, 2002, **60**, 279.

7 W. Dolata, E. Piotrowska, J. Wajdzik and J.Tritt-Goc, *Meat Science*, 2004, **67**, 25.

8 M.A. Cabrera, P. Gou, L. Foucat, J.P. Renou and J.D.Daudin, *Meat Science*, 2004, **67**,169.

9 T.M. Guiheneuf, A.D. Parker, J.J Tessier and L.D. Hall, *Magnetic Resonance in Chemistry*, 1997, **35**, S112.

10 R.S. Kirk. *Pearson's Composition and Analysis of Foods*. Harlow, Longman, 1991.

11 H.C. Bertram, S.B. Engelsen, H. Busk, A.H. Karlsson and H.J. Andersen, *Meat Science*, 2004, **66**,437.

12 E. Micklander, B. Peshlov, P.P. Purslow and S.B. Engelsen, *Trends in Food Science & Technology*, 2002,**13**,341.

13 A.J. Fowler and A. Bejan. *Int. J. Heat and Fluid Flow*, 1991, **12**, 375-383.

14 H.R. Cross, P.R. Durland and S.C. Sideman, 'Sensory Qualities of Meat' in *Muscle As Food*, ed. P.J. Bechtel, Academic Press, London, 1983. Chapter 7, pp.279-320.

15 G. Offer, D. Restall and J. Tinick, 'Water-Holding in Meat' in *Recent Advances in the Chemistry of Meat*, ed. A.J. Bailey, The Royal Society of Chemistry,London,1983. pp 71-86.

MRI AND NMR SPECTROSCOPY STUDY OF POST-HARVEST MATURING
COCONUT

Nikolaus Nestle[1], Arthur Wunderlich[2], Reinhard Meusinger[3]

[1]TU Darmstadt, Institut für Festkörperphysik, Hochschulstraße 6, D-64289 Darmstadt
[2]Universität Ulm, Abteilung Diagnostische Radiologie, D-89069 Ulm
[3]TU Darmstadt, NMR-Abteilung, Fachbereich Chemie, Petersenstrasse 20, D-64287
Darmstadt

1 INTRODUCTION

The coconut (Cocos nucifera) is a major agricultural product of many tropical regions of
the world. It is consumed both in the green stage (where the coconut water provides an
excellent refreshing beverage) and in the mature stage (when the fibrous, fatty pulp is the
product of interest). Dried coconut pulp is used for the extraction of its oil which is then
used for nutritional, cosmetic and other purposes. Despite its economic importance and its
convenient size for MRI studies in clinical MR systems, only very few MRI studies on
coconuts can be found in the literature[1,2]. In both cases, only images obtained on mature
coconuts were published. Furthermore, only cursory remarks on the structure and
contrasting behaviour of the pulp layer in coconuts can be found in Ref. 1, and no study of
the post-harvest maturation process on one coconut is known to us.

In addition to the lack of data on the structural changes during the maturation process in
the coconut, especially green coconuts provide nice features as a test sample for exploring
and development of new contrast mechanisms in MRI under the action of internal
magnetic field gradients:
- o The fibrous mesocarp tissue exhibits quite strong effects of internal magnetic field
  gradients which are typical for plant tissues with air enclosures[3].
- o In the endosperm (pulp), we observe indications for similar effects of magnetic
  inhomogeneity on an even smaller length scale while
- o The liquid in the lumen of the nut provides an internal reference material.

2 MATERIALS, METHODS AND RESULTS

2.1 Coconut specimens

Three charges of green coconuts were purchased during the work. A first set of fruits was
purchased by one of the authors near Atacames/Ecuador and transported to Germany as
part of normal travelling luggage. A delivery of additional (of Caribbean origin) was

obtained from a commercial exotic fruit dealer (Fruchthof Nagel, Neu-Ulm, Germany). For the Ecuadorian samples, the time between harvesting and the first measurements was three days. For the other samples, it is not exactly known to us but probably similar or even a little longer. Between the experiments, the drupes (botanically speaking, the coconut is no nut but a drupe) were stored at ambient conditions in an air-conditioned office building (at temperatures of about 22 °C and a relative moisture content of about 30 %) without further measures for moisture and temperature control. Spectroscopy was performed on sample materials taken from two coconuts of Caribbean origin which were selected on the basis of the MRI results. Furthermore, a specimen of brown coconut (unknown origin and storage history) obtained at a local supermarket was studied by MRI, too.

2.2 NMR equipment used

MRI was performed on a standard Siemens Magnetom Vision clinical MRI scanner (Siemens Medical, Erlangen, Germany) operating at a magnetic field of 1.5 T, corresponding to a proton resonance frequency of 63 MHz. Postprocessing of the images was performed by a combination of IDL macros (conversion of the Siemens data to 3D analyze), MRIcro[4] (a freeware-package available from the University of Nottingham), and home-written FreePascal programs.

NMR spectroscopy was performed on a Bruker Bruker Avance DRX 500 spectrometer. Static ^1H, ^{13}C and HSQC (Heteronuclear Single Quantum Correlation) 2D experiments were performed on the liquid samples. On the solid samples, magic angle spinning at 4.2 kHz was performed to record ^1H and ^{13}C 1D spectra. All ^1H spectra were acquired using a water suppression by presaturation.

2.3 MRI results

2.3.1 MRI sequences applied

All coconuts were examined using a range of different imaging protocols with different parameter weighting in order to be able to extract unambiguous information on the contributions of spin density, relaxation and diffusive phenomena to the resulting signal intensities recorded in the imaging experiments.

3D-FLASH images with an echo time of 1.8 ms and a repetition rate of 4.4 ms were acquired at different flip angles ranging from 1 ° to 30° in order to achieve different degrees of saturation. For flip angles of 5° and more, a good-quality dataset with 64×160×256 data points (and a spatial resolution of (2.5mm)×(1mm)×(1mm)) could be acquired on 3 minutes. For a flip angle of 1° (which comes closest to a spin density image of all sequences used in this study), more data accumulation was necessary due to the low signal intensity, so these datasets took at least 6 minutes to acquire. In the further course of this article, the FLASH data shall be presented using the terminology FL-A with A indicating the flip angle in degrees.

Multislice spin echo images were performed with echo times ranging from 17 ms to 256 ms and repetition times between 1000 ms and 9999 ms. The slice thickness in the images was either 1 mm or 2 mm. Like in the FLASH images, a rectangular field of view was sampled. Furthermore, a data matrix reduction to 70 % in the phase encoding direction was applied. The number of data points in the read gradient direction was typically 256 data points corresponding to an in-plane pixel size of about (1mm)×(1.4mm). For images with a repetition time of 9999 ms, the data matrix size was reduced to 128 points in read gradient direction. In the further course of this article, spin-echo-data shall be referred to as SE-TE-TR with TE indicating the echo time in ms and TR indicating the repetition time in ms. Multislice turbo spin echo images were acquired using echo times of 54 ms, 96 ms and 256 ms. The corresponding repetition times were 3000 ms, 3300 ms and 9999 ms, respectively. The so-called turbo factor (i.e. the number of echoes acquired in one echo train) for the images was 5, 7 or 13, respectively. Turbo spin echo data shall be referred to as TSE-TE-TR in the further course of the article. At the same nominal echo times, turbo spin echo and spin echo data exhibit quite different sensitivity to the effects of internal magnetic field gradients[3] as the additional refocusing pulses applied in the turbo spin echo lead to a reset in the diffusive signal attenuation at the formation of each echo (like in a CPMG-pulse train for NMR relaxometry).

2.3.2 Typical MRI results

2.3.2.1 Brown coconut

In Figure 1, images obtained on the brown coconut sample are presented. Before imaging, the coconut shell was moistened in order to impart some NMR visibility also to the shell.

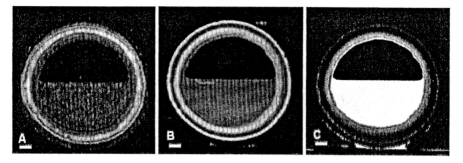

Figure 1 *Transversally oriented MRI scans of a mature brown coconut. A: strongly T_1-weighted FL-30 image, B: strongly T_1-weighted SE-12-352 image, T_2-weighted TSE-96-5000 image. The bright bar corresponds to 1 cm.*

The comparison of the images reveals a complex structure of the pulp layer. Starting from the outside, we find a thin water-rich-layer on the interface between the shell and the endocarp which is probably due to water ingress from the wetting of the shell. The outermost endosperm layer consists mainly of material with a very short T_2 which is not even visible in the FL-30 image. The T_2 of the endosperm progressively increases from the outside to the inside. At the same time, also T_1 increases. As a consequence of this, we find the regions with the highest signal intensity in the FL-30 image quite close to the shell while the corresponding region in the SE-12-352 is a little bit more to the inside (where the T_2 values are already sufficiently long). Coming closer to the inside of the nut, we find material with even longer relaxation times that even provides some signal intensity in the TSE-96-5000. Finally, at the innermost layer of the endosperm we find a water-rich gel phase. In the SE-12-352, we can se some chemical shift artifact leading shifting the signal of the fatty components of the endosperm layers to the right. As there is no similar shift artifact in the TSE-96-5000 (despite even lower read gradient strengths), we can infer that the signal from the endosperm in Figure 1C is only due to water droplets in the fatty endosperm matrix and not due to long-T_2 fat components.

The observation of a rich internal structure in the endosperm tissue of the brown coconut provided the motivation for further studies on green coconut specimens in order to identify the way the structure of the endosperm is formed with time.

2.3.2.2 Green coconut

Figures 2 and 3 show typical imaging results obtained on green coconuts (both coming from the same delivery of Caribbean origin). From the outside appearance (size, shape, uniformity and hue of colour, etc.), no clear indications could be found for the amount of pulp to be expected inside the individual nuts. However, there seems to be a correlation between the presence of a thick pulp layer and a low signal intensity recorded from the mesocarp layer on the stem side of the coconut.

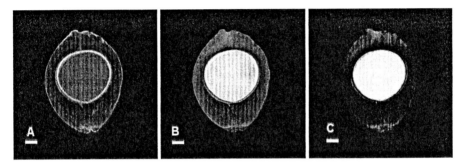

Figure 2 *Sagitally oriented MRI scans of a freshly purchased green coconut with very thin pulp layer. A: T_1-weighted FL-15 image, B: moderately T_1-weighted SE-17-2000 image, T_2-weighted TSE-96-3300 image. The bright bar corresponds to 2 cm.*

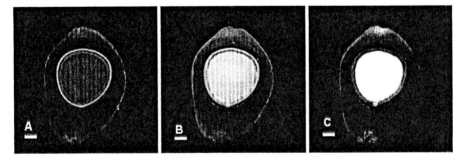

Figure 3 *Sagitally oriented MRI scans of a freshly purchased green coconut with very thin pulp layer. A: T_1-weighted FL-15 image, B: moderately T_1-weighted SE-17-2000 image, T_2-weighted TSE-96-3300 image. Note the embryo already formed in the endosperm (most obvious in image C). The bright bar corresponds to 2 cm.*

Despite the great variability in the thickness of the pulp layer, further features of the drupe's anatomy are observed in quite similar manner in all specimens studied:

- o The pericarp (with some content in fatty substances which make the fruit water-proof) sticks out with enhanced signal intensity in the FLASH measurements,
- o The longer the echo time is, the more obviously some large vials with a high water-content can be distinguished from the surrounding mesocarp tissue; some especially large of those vials connect the stem and the reminders of the pistil with the endocarp while the majority of the vials surrounds the endocarp in a parallel fashion.
- o In sequences with short and intermediate echo times, considerable signal intensity is recorded from the endocarp tissue (i.e. from the developing inner shell of the coconut). The shape of the endocarp and the position of the "eyes" in the nut can be clearly identified from the images.

In addition to ordinary spin echo images, also fat-saturated images (taking advantage of the standard binomial fat saturation routine of the Magnetom) were acquired for some coconut specimens. In those images, no change in the contrast between the pulp and the coconut water was found for early stages of pulp development such as in Figure 2. For coconuts with further developed pulp (such as in Figure 3), a very pronounced change in the signal intensity from the pulp layer was observed due to fat saturation.

2.3.3   *Post-harvest changes in green coconut specimens*

As indicated in the discussion of the images obtained on the brown coconut, one main intention of the study was to observe the formation of the inner structure of the endosperm tissue during the maturation of the coconut. Images obtained on the sample with the originally thin endosperm layer are given in Figure 4. The growth in thickness of the endosperm layer is clearly visible from the images. Furthermore, the longitudinal relaxation rate in the pulp layer shows a strong decrease as can be seen from the plot of the contrast intensity as a function of the excitation angle in the FLASH images (see Figure 5).

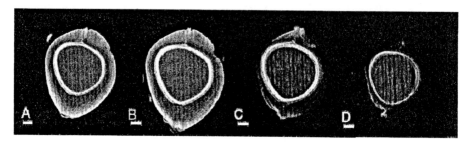

Figure 4   *FL-15 images of the same green coconut A: about 3 days post-harvest, B: about 10 days post-harvest, C: about 6 weeks post-harvest about 9 weeks post-harvest. The bright bar corresponds to 2 cm.*

After a storage time of two months, the coconut started to lose a considerable amount of water (an exit point where the water left the endosperm could not be unambiguously identified from the images). At the same time, the contrast between the endosperm and the remaining water decreased dramatically. This increase in the longitudinal relaxation time can be attributed to a decomposition of the fatty components in the endosperm. In a destructive analysis of the fruit done at this stage, a structureless and very soft endosperm tissue was found. From the sensory appearance of the fruit, it was severely rotten and no further analysis of the water or the endosperm was performed.

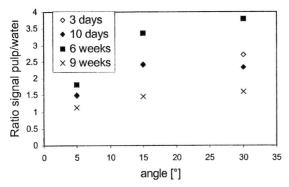

Figure 5 *Ratio of signal intensities of pulp and water as a function of the excitation angle in FLASH images of the coconut shown in Figure 4. Note especially the drop in contrast after 9 weeks.*

Figure 6 shows images obtained on a coconut specimen with a thick endosperm layer and an already visibly embryo, which was stored on a moist sand-bed in order to observe the further development of the embryo.

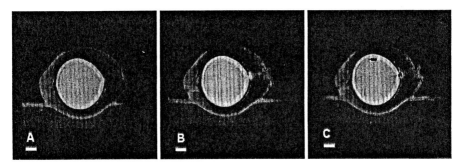

Figure 6 *SE-17-2000 images of a green coconut with a thick pulp layer and a clearly visible embryo A: about 5 days post-harvest, B: about 2 weeks post-harvest, C: about 3 weeks post-harvest. The bright bar corresponds to 2 cm. The fruit was stored on a moist sandbed in order to establish favourable conditions for germination.*

In contrast to the sample with a thin pulp layer shown in Figure 4, there is only a minor increase in the thickness of the pulp layer here. Furthermore, no loss of internal structure of the pulp takes place upon loss of part of the water in this case. In the germinating sample, the loss of water occurs through the hole created by the embryo penetrating the endocarp. In all the three coconuts where we could clearly identify an embryo, some structure in the outgrowing embryo could be observed but the process stopped before the formation of a haustorium dissolving the endosperm[5] and so no further development of a viable seedling could be observed. The most probable explanation for our failure to observe the further steps of the germination process is the low relative moisture in the room where the germinating coconuts were stored between the measurements. Furthermore, the state of maturity of the green coconuts at the point of harvesting may also have been suboptimal for successful germination.

2.4 NMR spectroscopy results

The coconut samples shown in Figures 2 and 3 were opened after the imaging experiments in order to take water and pulp samples for spectroscopy. In Figure 7, the spectroscopic results obtained on the materials harvested from the coconut of Figure 2 are presented. Spectra of the material sampled from the specimen in Figure 3 are shown in Figure 8. The liquid spectra and the spectrum of the emerging pulp layer are dominated by saccharides and some amino acids, while the spectrum of the thick pulp is completely dominated by fatty acids.

Comparing our results to those presented in Ref. 1, we can unambiguously assign the peak at 1.1 ppm to the methyl group of ethanol (which is a quite common metabolic intermediate in the maturation of many fruits) as no sufficiently strong CH_2-signal from the chains of fatty acids is found along with it in the water spectrum shown in Figure 7. The identification of the peak as ethanol was confirmed by the [13]C (lines at 17 and 58 ppm) and HSQC results.

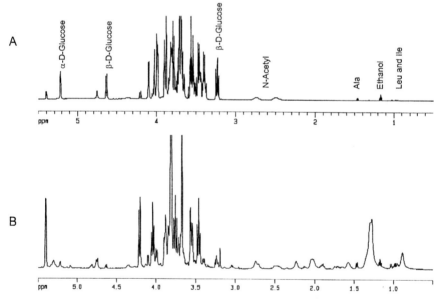

Figure 7   *^{1}H NMR spectroscopy results of (A) water and (B) pulp sampled from the coconut specimen of Figure 2. The main features in the spectra around 4.3 to 3.2 ppm are due to several carbohydrates (in addition to glucose also galactose and saccharose) and different amino acids (other lines of alanine, leucine and isoleucine annotated in the spectra). The pulp spectrum furthermore shows some signal stemming from fatty acids.*

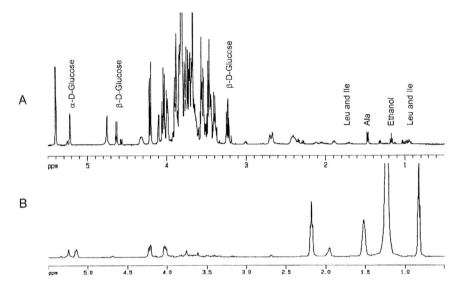

Figure 8 *¹H spectra obtained on (A) water and (B) pulp sampled from the coconut specimen in Figure 3. The water spectrum is again dominated by various sugars and amino acids (with different intensities in the sugar pattern); the pulp spectrum now is almost entirely due to (mostly saturated) fatty acids.*

3 CONCLUSIONS AND PATHWAYS FOR FURTHER RESEARCH

The MRI results indicate that the development of the internal structure of the pulp layer in a maturing coconut is a complex phenomenon that includes several stages. In specimens of green coconut with initially only very thin pulp layers, both a growth in thickness of the endosperm layer and a change to more fatty composition could be clearly observed. The internal layered structure in the green specimens nevertheless was always different from that observed in the brown coconut. Comparisons between freshly harvested coconuts with differently developed pulp layers indicate that the water distribution within the mesocarp tissue is different in fruits that already exhibit a thick pulp layer. This observation may allow the development of a non-destructive sensing strategy for the pulp development by simple non-NMR sensors. In the NMR spectra, we could identify ethanol as a component present in the coconut water. A more in-depth analysis of the spectra is beyond the space restrictions in this contribution and shall be presented in a separate paper.

References

1 N.R. Jagannathan, V. Govindaraju and P. Ragunathan *Magnetic Resonance Imaging* 1995, **13**, 885.
2 F. Schick *Magnetic Resonance Imaging,* 2001, **17**, 527.
3 N. Nestle, A Qadan, P. Galvosas, W. Süss and J. Kärger *Magnetic Resonance Imaging*, **20**, 567.
4 C. Rorden and M. Brett, *Behavioural Neurology* 2000, **12**, 191.
5 JL Verdeil, V. Houcher, *Trends in Plant Science* 2002, **7**, 280.

ANALYSIS OF BLENDS OF RAW COFFEES OF ARABICA AND ROBUSTA VARIETIES BY ^{1}H NMR AND CHEMOMETRIC METHODS

L.A.Tavares,[1] A.G. Ferreira,[1] M.M.C. Ferreira,[2] A. Correa[3] and L.H. Mattoso[3]

[1] Departamento de Química, Universidade Federal de São Carlos, São Carlos - SP, BR
[2] Instituto de Química, Universidade Estadual de Campinas, Campinas - SP, BR
[3] Embrapa Instrumentação, São Carlos - SP, BR

1 INTRODUCTION

The coffee drink is prepared from arabica or robusta specie grain, or even from blends of them. Arabica coffee presents more pronounced and refined flavor and its price is higher than robusta.[1-3] Although the best coffee drink is obtained from arabica specie grain, blends of both are necessary in several proportions depending on consumers taste, the price rate or if it will be used to produce instant coffee or not, because robusta specie yields more soluble solids than arabica.[4] Although arabica and robusta can been distinguished by their grain size, this visual criterion is not practical for blends with inhomogeneous grain size and it is useless after roasting and grinding. Due to this and many other reasons there is a great need for techniques that can discriminate both species and help the taster in the coffee classification to maintain the blends quality.

Although we have several analytical techniques like: Mass Spectrometry[5], High Performance Liquid Chromatography[6-8], Infra Red Spectroscopy[9-10], Gas Chromatography[11-12], etc which are applied in the investigation of some molecular components, all of these methodologies, to some extent, require some extraction process. ^{1}H NMR spectroscopy has the advantage of not requiring any pretreatment when compared to all the other techniques and, moreover, that a single ^{1}H NMR spectrum is able to provide valuable information about the structure and composition of the main compounds present in the sample. The only extraction procedure was a single boiling water extraction using an expresso coffee maker. Using this approach we simplified the analysis and minimized the possible qualitative and quantitative compound modifications from the original mixture, which can be originated from exhaustive extraction and/or purification methods and chemical derivations.[13] The other advantage is that with a minimum sample preparation we can make the analysis quickly with a maximum number of samples, a requirement for authenticity and quality control analysis. However, without appropriate methods of data analysis the spectroscopic details which potentially make these techniques so powerful would become overwhelming. To overcome this problem chemometric methods are now being used directly on NMR spectral data for classification and discrimination purposes. [14-16] In this work chemometric methods were applied to ^{1}H NMR spectroscopy data to classify green coffee grains of arabica and robusta blends into two groups, one with samples that contain more arabica coffee and the other robusta (classificatory analysis), and also to determine the percentage of each specie into the blends.

The following chemometric methods were used[17-20]: Principal Component Analysis (PCA) and Hierarchical Clusters Analysis (HCA) for exploratory data analysis, k-Nearest Neighbours (KNN) and Soft Independent Modelling of Class Analogies (SIMCA) for classificatory analysis and, Principal Component Regression (PCR) and Partial Least Squares (PLS) for quantification.

2 MATERIALS AND METHODS

2.1 Sample Preparation

The samples, collected in Southern Minas Gerais, were prepared from ground green coffee of arabica and robusta species in proportions according to Table 1. The "coffee" was extracted with an expresso coffee maker with 72.0 mL of boiling water and 10.0 g of ground coffee. All the spectra were obtained in triplicate for each sample (indicated by the letters a, b and c) using 0.6 mL of coffee extract with three drops of D_2O.

Table 1   *Composition in percentage of arabica and robusta in coffee samples*

Sample	% Arabica	% Robusta
A0	0	100
A1	10	90
A2	20	80
A25	25	75
A3	30	70
A35	35	65
A4	40	60
A5	50	50
A6	60	40
A65	65	35
A7	70	30
A75	75	25
A8	80	20
A9	90	10
A10	100	0

2.2   ¹H NMR Spectra

All the ¹H NMR spectra were carried out using a Bruker DRX-400 spectrometer equipped with a 5 mm inverse probehead, maintaining the temperature constant at 298K and the acquired processing data were measured with the same parameters. Water suppression was achieved using the 'zgcppr' pulse sequence which uses composite pulses with different phases better than a single 90° pulse. The data were acquired with 64 FIDs, acquisition time 4.0s, spectral width 8119 Hz, irradiation power for water suppression 55dB, recycle delay 2.0s and processed with zero-filling using a line broadening 0.3. The phase and baseline correction were manual setup using the Bruker software.

2.3   Chemometric Analysis

The Pirouette® software package was used for data analysis. The original data matrix was constructed from the spectra of 45 objects, each spectrum described by 2100 points (variables). Principal Components Analysis was applied for optimisation and feature selection resulting in a subset (45 x 730) which contains basically the main information. Different methods of multivariate analysis were applied to this new data matrix. PCA and HCA for an exploration, KNN and SIMCA for classification, and PLS and PCR for quantification analyses[15-18]. Before building the models, each spectrum was normalized to set the maximum signal at 100 and first derivative was taken. In the next step, the data was autoscaled, in which each column is mean-centered and then divided by its standard deviation. For HCA, incremental linkage method was chosen and the maximum of 10 nearest neighbours was used for KNN calculations.

3   RESULTS AND DISCUSSION

The preliminary data processing such as: calibration, phasing and baseline correction was carried out on the ^{1}H NMR spectra to obtain all the same processing spectrum as possible in order to reduce discriminant effects originated by processing variations, and to evidence the differences between sample compositions. The spectrum is shown in Figure 1. If there are differences between spectra derived by different sample compositions or by processing, the chemometric analysis makes discriminations between them and will classify samples in different clusters.

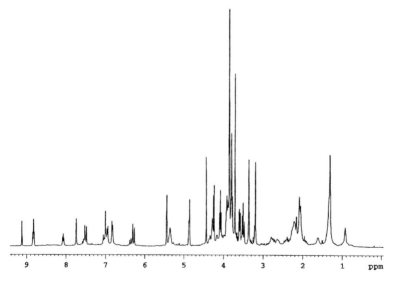

Figure 1   *The ^{1}H NMR spectra used in the analysis*

PCA was mainly used as a guide to help in choosing which variables are most important to discriminate the samples and to select the appropriate preprocessing and data transformation to be applied to the data matrix. This methodology makes it possible to identify the most important directions of variation in a multivariate data matrix and to present the results in intuitive graphic plots. The aromatic region of spectra, without noise background, gives the best sample discrimination. However, an accurate examination of the loadings combined with the original spectra was not enough to understand the basis of the clustering behaviour, but we believe that chlorogenic acids and caffeine are most important for this discrimination. The reason for this is that both of these substances are in higher concentration in robusta coffee. This can be seen in Figure 2 which shows five groups of signals from 6.0 to 7.8 ppm, representing an increase in concentration in the robusta specie.

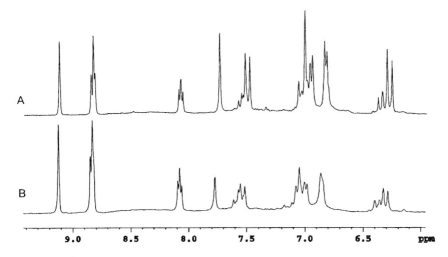

Figure 2 *^1H NMR spectra showing the region of aromatic compounds used in the chemometric analysis. A) robusta and B) arabica*

Principal component analysis (PCA), applied to the data matrix containing only the signals from the aromatic region of ^1H NMR spectra, shows that it is possible to discriminate samples into two groups. One group has a high percentage of arabica specie, on the right side (●), and the other more robusta, on the left (■). This can be observed in score plot of PC1 x PC2, Figure 3.

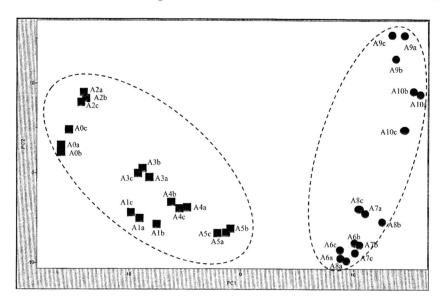

Figure 3   *Score plot (PC1(62.25%) x PC2 (17.85%)) showing the discrimination between samples from arabica (●) and robusta specie (■)*

    The dendrogram obtained from HCA (Figure 4) is in agreement with PCA results. With a similarity index of 0.45 the samples are clustered into two main groups. Moreover, in HCA the results present a behaviour trend to grouping the samples in the same direction when we increase the amount of one component. In this dendogram, like as in the PCA, we can notice that there are two main groups and two subgroups. One main group without or little arabica content (subgroups A0, A1 and A2) and another one with increased arabica content (subgroups A3, A4 and A5). The other main group include samples which have 50% or more percentage of arabica specie, including one subgroup with approx. 100% arabica (subgroups A9 and A10). We also observe that samples with less content of arabica coffee are located in the lower part and as the the amount of Arabica increases the samples appear higher up in the plot. For instance, samples A0a, A0b and A0c don't have arabica specie and A10a, A10b and A10c are 100% arabica. Samples containing 50% arabica and 50% robusta (A5a, A5b and A5c) are located between the two groups.

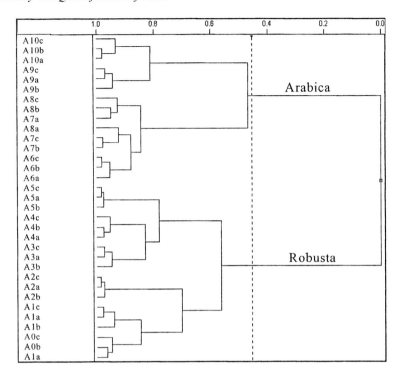

Figure 4 *Dendrogram obtained from HCA for green coffee samples*

The samples A2a-c, A25a-c, A35a-c, A65a-c, A75a-c and A8a-c (a, b and c are triplicates) were used to validate the classification and calibration models, and the others mencioned in Table 1 were used for building the models. For classification models KNN and SIMCA, these samples were predicted correctly, except for samples A65a-c, which were wrongly classified in the group of samples with high content of robusta specie. The KNN model appeared to be more efficient than the SIMCA model, since it didn't classify any sample in the wrong group when using from 1 to 10 nearest neighbours, and furthermore correctly classified fifteen of eighteen test samples (Table 2). SIMCA could classify only two samples in the correct group as shown in Table 2, but all the test samples (except A65a-c) were located near to their own class. The samples from A0 to A5 near to class 1 (robusta) and A6 to A10 near to class 2 (arabica), figure 6. The samples A5a-c (50/50 arabica/robusta) were used to build the models.

Table 2   *Class prediction obtained by KNN and SIMCA models*

Sample	KNN predict	SIMCA predict
A2a	1	-
A2b	1	-
A2c	1	-
A25a	1	-
A25b	1	-
A25c	1	-
A35a	1	-
A35b	1	-
A35c	1	-
A65a	1	-
A65b	1	-
A65c	1	-
A75a	2	2
A75b	2	-
A75c	2	-
A8a	2	-
A8b	2	2
A8c	2	-

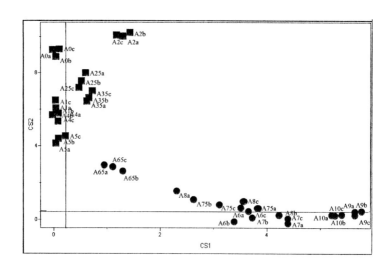

Figure 6   *Class prediction obtained by the SIMCA model (■ samples near to group 1, ●*
samples near to group2)

The PLS and PCR regression models were built on 27 samples and leave one out cross-validation was used to determine the optimum number of factors, which was found to be 5, giving a correlation coefficient of 0.996 and 0.993, respectively.

The proportion of each coffee species into the blend was then predicted using the PLS and PCR models using an external validation test set. PLS was more accurate than PCR when we compared the predicted with the experimental values (Table 3). When we try to predict the content of one species that is close to the other, the models are not so good.

Table 3 *Values of arabica content into the coffee blend predicted by models PLS and PCR*

Sample	PLS predict	PCR predict	Experimental Value for arabica (g)
A2a	1.9	2.5	2.0
A2b	2.0	2.6	2.0
A2c	1.8	2.5	2.0
A25a	2.1	2.2	2.5
A25b	2.0	2.3	2.5
A25c	2.2	2.5	2.5
A35a	3.4	3.6	3.5
A35b	3.3	3.5	3.5
A35c	3.4	3.7	3.5
A65a	6.5	6.9	6.5
A65b	6.3	6.7	6.5
A65c	6.2	6.7	6.5
A75a	7.7	7.7	7.5
A75b	7.8	7.9	7.5
A75c	7.6	7.7	7.5
A8a	8.2	8.3	8,0
A8b	8.3	8.3	8,0
A8c	7.7	8.0	8,0

Roasted coffee was also studied using the same procedure as for green coffee, but the results were not so promising. One of the reasons could be that the level of roasting can change the final composition of each blend and consequently change the data analysis results.

4 CONCLUSIONS

The results presented show that ^1H NMR spectroscopy together with the chemometric methods can be used as a rapid method to identify and quantify the arabica and robusta contents in green coffee blends. The analysis was based mainly on the differences of aromatic compound contents, especially chlorogenic acids and caffeine.

References

1   A.B. Joly, *Botânica Econômica - As principais culturas Brasileira*, Ed. Hucitec-EDUSP, 1978.
2   R. Briandet, E.K. Kemsley and R.H. Wilson, *J. Agric. Food Chem,* 1996, **44**, 170.
3   F. Carrera, M. León-Camacho, F. Pablos and A.G. González, *Anal. Chim. Acta,* 1998, **370**, 131.
4   V.D. Carvalho, S.J.R. Chagas and S.M. Chalfoun, *Inform. Agropec,* 1997, **18**, 5.
5   L. Maeztu, C. Sanz, S. Andueza, M. P. De Peña, J. Bello and C. Cid, *J. Agric. Food Chem*, 2001, **49**, 5437
6   M.L.R. Del Castillo, M. Herraiz and G. Blanch, *J. Agric. Food Chem.*, 1999, **47**, 1525.
7   M.J. Martin, F. Pablos, and A.G. González, *Talanta*, 1998, **46**, 1259.
8   J.F. Cotte, H. Casabianca, B. Giroud, M. Albert, J. Lheritier, M.F. and Grenier-Loustalot, *Anal Bioanal Chem*, 2004, **378**, 1342.
9   Z. Bouhsain, J. M Garrigues, S. Garrigues and M. Guardia, *Vibrat Spec*, 1999, **21**, 143.
10 L. Pillonel, W. Luginbuhl, D. Picque, E. Schaller, R. Tabacchi and J.O. Booset, *Eur Food Res Technol.*, 2003, **216**, 174.
11 C.P. Bicchi, O.M. Panero, G.M Pellegrino, and A.C Vanni, *J. Agric. Food. Chem.* 1997, **45**, 4680.
12 F. Carrera, M. León-Camacho, F. Pablos and A.G. González, *Anal Chim Acta*, 1998, **370**, 131.
13 M. Bosco. R. Toffanin, D. De Palo, L. Zatti and A. Segre, *J. Sci. Food Agric*, 1999, **79**, 869.
14 J.W.E. Vogels,. L. Terwel, A.C. Tas, F. Van Der Berg, F. Dukel, and Van Der Greef. *J. Agric. Food Chem.*, 1996, **44**, 175.
15 P.S. Belton, I.J. Colquhon, E.K. Kemsley, I. Delgadillo, P. Roma, M.J. Dennis, M. Sharman, E. Holmes, J.K. Nicholson, and M. Spraul, *Food Chem.*, 1998, **61**, 207.
16 I.F. Duarte, A. Barros, C. Almeida, M. Spraul and A.M. Gil, *J Agric. Food Chem.*, 2004, **52**, 1031.
17 H. Martens and T. Naes. *Multivariate Calibration*. John Wiley & Sons. 1989.
18 K. Beebe, R. Pell and M.B. Seasholtz. *Chemometrics – A Practical Guide*. John Wiley & Sons. 1998.
19 R. G. Brereton. *Chemometrics Data Analysis fro the Laboratory and Chemical Plant*. John Wiley & Sons. Chichester. 2002.
20 D.L.. Massart, B.G.M. Vandeginste, L.M.C. Buydens, S. De Jong, P.J. Lewi and J. Smeyers-Verbeke. Handbook of Chemometrics and Qualimetrics: Data Handling in Science and Technology. Volumes 20A and B. Elsevier. Amsterdam. 1997.

Functionality, Structure and Ingredients

USE OF STATE-OF-THE-ART NMR IN BEER PRODUCTION AND
CHARACTERIZATION

Jens Ø. Duus

Carlsberg Laboratory, Gamle Carlsberg Vej 10, DK-2500 Valby, Denmark

1 INTRODUCTION

The paper reviews several applications of NMR from research projects towards beer and
the production of beer carried out in the Carlsberg Laboratory. It is a general theme that a
better understanding of the fundamental structure of the component in the beer will aid the
search for solutions of process problems and properties of beer during production and
storage.
The structure determination of beer component from small natural compounds over large
polysaccharide to the complex mixture of all of these has mainly been based on NMR with
aid of mass spectrometry.

The beer production, as such, will not be described here, but the paper will touch on
problems connected to a) wort filtration, b) composition of the hops and c) final beer.

2 RESULTS

2.1 Nano-probe NMR of arabinoxylans oligosaccharides

Beta glucan is an important constituent in the barley cell wall and is a linear
polysaccharide of glucopyranose linked by β-(1-3) and β-(1-4) bonds. It is well known that
barley beta-glucan, when insufficiently degraded by the natural enzymes, forms viscous
solutions and causes problems in the brewing process[1]. We have previously described the
use of NMR for the characterization of the enzymatic degradation of the polysaccharide [2,3].
More recently it was observed that even when the beta-glucan was well degraded, for
example, by adding external enzymes, problem still arise with long filtration times. This
has led us to look closer into the structure and enzymatic degradation of another cell wall
constituent in barley, the arabinoxylan [4-6].
Barley arabinoxylan consists of a β-(1-4)-D-xylopyranose backbone carrying O-2, O-3 or
O-2,3 α-L-arabinofuranosyl substituents. This polysaccharide is the main constituent of
both the aleurone and endosperm cell walls [7] and requires a combination of both
arabinofuranosidases and xylanases for efficient degradation.

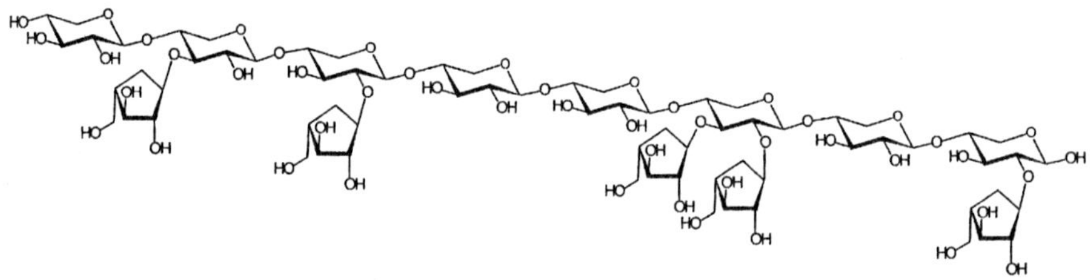

Figure 1 *A schematic representation of the arabinoxylan with the different types of mono and di-substituted residues.*

For a detailed description of a key enzyme in the degradation, the barley arabinofuranosidase, a series of smaller oligosaccharides were isolated[8]. From barley arabinoxylan a mixture of oligosaccharides were produced by treatment with a commercial enzyme mixture and fractionated by size-exclusion chromatography. This afforded mixtures of smaller oligosaccharides which needed further separation.

Traditionally, this would require the use of larger preparative schemes using, for example, preparative HPAEC separation. However, it could here be shown that the use of a Varian nano-probe[9] at 500 MHz would allow for acquisition of standard one- and two-dimensional ^{1}H spectra using material from an analytical HPAEC column. It was determined that less than 5 nanomole of pure oligosaccharide was required for a set of standard two-dimensional spectra, as TOCSY, ROESY and DQFCOSY.

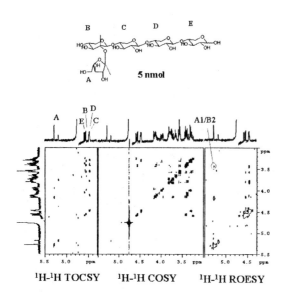

Figure 2 *500 MHz NMR data obtained on approx. 5 nmol of pentasaccahride*

The backbone of the arabinoxylan is degraded by xylanases, which cleave the β-(1-4) linkage between two consecutive xylopyranoses. These enzymes, however, have difficulties getting to the backbone due to the high degree of substitution by arabinofuranosides and therefore require an action by arabinofuranosidases. We have isolated such an enzyme directly from barley and wanted to determine the substrate preferences in order to evaluate if this can be a key enzyme in the natural degradion of the polysaccharide. Using the isolated oligosaccharides, it was then possible to follow the enzymatic degradation directly in NMR using a nanoprobe [10] and determine the specificity of the enzyme. Using NMR it is furthermore possible to determine if the enzymatic hydrolysis occurs via retention or inversion of configuration during the reaction.

2.2 Identification of reduced iso-α-acids derived from hops.

Almost any modern beer contains a form of hops (*Humulus lupulus*). The complex mixture of natural compounds from the hops is furthermore converted during the beer production and all together the many components add bitter taste and also contribute to the foam-forming properties of the beer[11,12].

The iso-α-acids formed during wort boiling are key components, but unfortunately susceptible to degradation, which can lead to unwanted taste components[13]. One way to circumvent this is by using synthetically reduced iso-α-acids.

We have used NMR to determine the structure of several of the components in the standard mixture of tetrahydro-iso-α-acids, hexahydro-iso-α-acids and rho-iso-α-acids after isolation using HPLC. As for many other natural compounds, the complete structure elucidation can be obtained by NMR if sufficient amounts of pure material can be isolated[14]. It was even possible to determine the relative stereochemistry of the side chain in the hexahydro-iso-α-acids. This was achieved by 4,6-isopropylidene derivatization of the two adjacent hydroxyl groups, which locks the rotation of the side chain and allows for an unambiguous assignment of the configuration of the 6-position using a combination of NOE and simple molecular modeling.

It can be seen how the restriction of the side chain rotation brings the H5 and H6 close in the 6S isomer and not in the 6R isomer, which can be substantiated by a NOE in the 6S isomer[14].

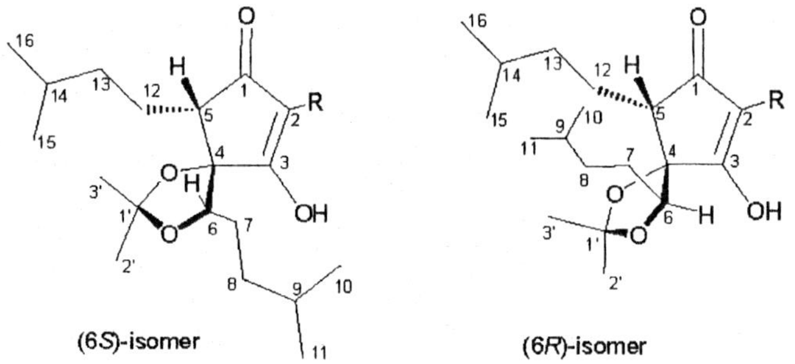

Figure 3 *Examples of the 6S and 6R isomers*

2.3 Quantification of components in beer by NMR

An obvious experiment having access to high field NMR and working with beer would be to acquire a simple one-dimensional NMR spectrum of beer. This can be done with very little sample preparation, essentially just adding a deuterated solvent for locking after degassing of the beer. The NMR experiment is likewise very simple. It just requires presaturation of the strong water signal and perhaps the ethanol signals; the latter, however, not necessary. A presaturation of the water signal is quite simple with modern instrumentation and affords a very informative spectrum.

In a simple one-dimensional spectrum of beer[15] one can identify several components simply by the chemical shifts and published values. This applies for several signals for the carbohydrate components, with signals corresponding to the starch fragments which have not been degraded during the brewing process. Likewise, several amino acids can be observed.

The use of LC-NMR/MS has furthermore proved very useful for further characterization[16,17]

We have extended the use of NMR of beer for the quantification of several amino and organic acids by the use of either simple integration of partial least squares methods[18]. Integration worked well for the acids having well-separated signals, and the use of PLS extended the number of components which could quantitated. In total, 6 organic acids and 12 amino acids could be quantified with good precision and down to few mg/L.

3 DISCUSSION AND CONCLUSION

It has been described how NMR can yield detailed information of structure and composition of importance to the understanding of the production of beer.

It can be projected that NMR still can contribute with information in many aspects of beer production. The technique has several major advantages, such as being able to analyse very complex mixtures without separation, as seen by the NMR of the beer itself. Currently, in the laboratory a main objective is the investigation of living material, such as brewers' yeast, using HR-MAS techniques.

The major limitation of NMR has been the lack of sensitivity, but new developments, such as cooled probe heads, will gradually increase the sensitivity. This can then further be combined with new chemometric methods to provide a very information-rich technique at a reasonable cost per sample.

Acknowledgements

All the members of the group and collaborators who contributed to the work reviewed and cited here are acknowledged for their dedicated efforts.

References

1. C. W. Bamforth, *Brewers Digest*, 1982, 22.

2. K. Bock, J. Ø. Duus, B. Norman, and S. Pedersen, *Carbohydrate Res.*, 1991, **211**, 219.

3. O. Olsen, K. K. Thomsen, J. Weber, J. Ø. Duus, I. Svendsen, C. Wegener, and D. von Wettstein, *Bio/Technology* , 1996, **14**, 71.

4. M. S. Izydorczyk and C. G. Biliaderis, *Carbohydr.Polym.*, 1995, **28**, 33.

5. R. J. Viëtor, S. A. G. F. Angelino, and A. G. J. Voragen, *J.Cereal Sci.*, 1992, **15**, 213.

6. J. Y. Han, *Food Chem.*, 2000, **70**, 131.

7. A. Bacic and B. A. Stone, *Aust.J.Chem.*, 1981, **8**, 475.

8. A. Broberg, K. K. Thomsen, and J. Ø. Duus, *Carbohydrate Res.*, 2000, **328**, 375.

9. P. A. Keifer, L. Baltusis, D. M. Rice, A. A. Tymiak, and J. N. Shoolery, *J.Magn.Reson., Ser.A*, 1996, **119**, 65.

10. H. Ferre, A. Broberg, J. Ø. Duus, and K. K. Thomsen, *Eur.J.Biochem.*, 2000, **267**, 6633.

11. M. Verzele, *J.Inst.Brew.*, 1986, **92**, 32.

12. M. Moir, *J.Am.Soc.Brew.Chem.*, 2000, **58**, 131.

13. C. S. Burns, A. Heyerick, D De Keukeleire, and M. D. E. Forbes, *Chem.Eur.J.*, 2001, **7** , 4554.

14. L.I. Nord, S.B. Sørensen, and J.Ø. Duus, *Magn.Reson.Chem.,* 2003, **41**, 660.

15. I. Duarte, A. Barros, P.S. Belton, R. Richelato, M. Spraul, E. Humpfer and A.M. Gil, *J.Agr.Food Chem.*, 2002, **50**, 2475

16. A.M. Gil, I.F. Duarte, M. Godejohann, U. Braumann, M. Maraschin and M. Spraul, *Anal.Chim.Acta*, 2003, **488**, 35.

17. I.F. Duarte, M. Godejohann, U. Braumann, M. Spraul and A.M. Gil, *J.Agr.Food Chem.*, 2003, **51**, 4847.

18. L.I. Nord, P. Vaag, and J.Ø. Duus, *Anal.Chem.,* 2004, **76**, 4790.

QUANTITATIVE MEASUREMENTS BY MRI OF FLOW VELOCITY AND MIXING INDEX IN A SINGLE-SCREW EXTRUDER

M. H. G. Amin[1], L. D. Hall[1], W. Wang[2] and S. Ablett[2]

[1] The Herchel Smith Laboratory for Medicinal Chemistry, University of Cambridge School for Clinical Medicine, Robinson Way, Cambridge, CB2 2PZ, UK.
[2] Unilever R & D Colworth, Colworth House, Sharnbrook, Bedford, MK44 1LQ, UK.

1 INTRODUCTION

It is well known that Magnetic Resonance Imaging (MRI) can be used to measure complex fluid flow through complex geometries[1-4]. For example, MRI has been applied to flow measurements in single-screw extruders[5-8] to assess mixing quality in repetitive-orifice flow[9] and in a scraped-surface heat exchanger[10].

Screw extruders are widely used in the food industry to pump and/or to mix a variety of products. Since measurement of the flow field and the quality of mixing in extruders is of fundamental industrial interest[11-14], current industrial practice mainly uses prior experience, computational fluid dynamics (CFD) modelling and measurements of specific operating parameters, such as temperature, pressure and torque for flow estimations; taking samples from the mixing line is used for measuring concentrations and then calculating indices of mixing[15-17].

For quantification of the quality of mixing, two statistical methods are used. One is "scale of mixing": a measure of the size of regions of segregation within the mixture which can be quantified by a length scale correlation function for immiscible mixing[11,15]. The second is the "intensity of mixing": a measure of divergence from the mean composition which can be used for any type of mixing[11,15]. Different definitions of "mixing index" are given in the literature[11,15-17], but the simplest and most commonly used is given as: the concentration variance of the mixture normalised to its maximum value at complete segregation.

This paper reports the use of MRI to measure quantitatively the flow velocities, and the goodness of mixing in a single-screw extruder of two streams of 1% aqueous sodium carboxymethylcellulose (CMC). These had been labelled with different amounts of manganese (II) ion so that their relaxation times were sufficiently different to create image contrast for quantification of their subsequent mixing.

2 MATERIALS AND METHODS

2.1 The Single-screw Extruder

The single-screw extruder was designed to fit inside a 31-cm horizontal bore magnet, shielded inside a radiofrequency (RF) Faraday cage[8]. The extruder was surrounded by a RF coil, and then mounted within the gradient set located in the bore of the magnet. All components in the Faraday cage are non-ferromagnetic; thus, the outer-body frame and extruder screw were made of PEEK; and the rotating shaft was supported by glass ball bearings in plastic races. The rotation of the extruder was controlled by a variable-speed DC motor and gear-box located outside the Faraday cage, 2 m from the nearer face of the magnet. A tachometer was used to monitor the rotation speed of the motor, which can provide shaft rotation speeds between 5 and 1200 revolutions per minute for time periods of over 10 hours.

The single screw is a double-flighted screw with a helix angle of 17.8^0 and 6 turns for each of the flights[8]. To provide sufficient signal-to-noise ratio (S/N) to support high quality MR images, the flight height is 0.5 cm and the channel width is 1.3 cm with a screw diameter (i.e. the barrel id) of 4 cm, length to diameter ration (L/D) of 4.575 and root diameter 3 cm. The MRI-RF coil can be located near the middle, near the die, or near the inlet regions. A second channel in the sample inlet allows two streams of fluids to be pumped through the system simultaneously and thereby allows mixing to be studied. The flow of the major fluid component (viscosity up to 10 Pa.s) was driven by a progressive cavity pump with a flow rate up to 36 litres per hour; that of the minor component was by a precision dispenser pump. Circulation of sample fluids was via reinforced PVC tubing (0.012m id). Hard nylon tubing (0.004m id) was used to connect the pressure gages (RS transducer 249-3959), 1 m away from the sample inlet and 2.4 m from the outlet of the extruder.

The 3D geometry of the screw extruder made it necessary to acquire each MRI scan at exactly the same point in each revolution; this was achieved using an optically gated apparatus, built in house. This gating ensured that the MR imaging repetition time was the time taken for one revolution.

2.2 The Fluids

Viscous fluids, based on 1% (w/w) aqueous sodium carboxymethylcellulose (CMC, Blanose 7H5SCF), were chosen to ensure laminar flow in the extruder. The fluids were made by first stirring the CMC powder in aqueous $MnCl_2$ at room temperature for 30 minutes; the dispersions were then stirred at 80^0 C for 30 minutes; finally the resultant solutions were left overnight at room temperature to "de-air".

To induce MRI contrast for the MRI measurements of mixing, the major flow component was made from 0.1 mmol/l $MnCl_2$ solution, whereas the minor component was from 1.6 mmol/l $MnCl_2$. At 2 Tesla their spin-lattice (T_1) relaxation times are 345 and 26 ms, and spin-spin (T_2) relaxation times are 122 and 8 ms respectively. The T_1 and T_2 values of the major component and hence the concentration of $MnCl_2$ were chosen to ensure a high signal level in the MR images of flow measurements and mixing, whereas the values for the minor component were selected so that its signals were slightly higher than the noise level to facilitate the analysis of mixing images.

A thermostatted cone-and-plane rheometer (Carrimed, model CSL 100) with a 2^0 cone angle and 4 cm diameter was used to measure rheological properties of the fluid. The power-law equation, shear stress $\sigma = \eta\dot{\gamma} = k\dot{\gamma}^n$, was used for the shear rate-shear stress relationship, where $\dot{\gamma}$ is the shear rate and η the shear viscosity[18]. For the 1% CMC (7H5SCF) solution, the measured value of the power-law index (n) was 0.44 and the consistency coefficient (k) was 4.90 (Pa s^n).

2.3 MRI Experimentals

The MRI scanner was a 31-cm-bore, 2 Tesla superconducting magnet (Oxford Instruments), connected to a Bruker MSL 100 console with TOMIKON operating software. The magnetic field gradients were provided by a home-built set of coils (0.11 m i.d.), with each axis driven by a pair of amplifiers (Tecron 7560 and 7570), which produce gradient fields up to 19.5 Gause/cm, with a slew rate of 125 μs. The radiofrequency (RF) coil was an eight-strut-quadrature-birdcage coil (9 cm length, 7.9 cm i.d.) built in house. MRI data were processed in a network of UNIX workstations; the C-code software, required for processing and visualisation of imaging data, was developed in house.

The MRI measurements were conducted at 21^0C with either one or two fluids (for flow or mixing measurements, respectively) pumped through the extruder while the screw was rotating at either 20 or 30 rpm. The RF probe was positioned near either the inlet or outlet of the extruder (which are located *ca.* 6 cm from the centre of the images).

A gradient-echo (GE) pulsed-field-gradient (PFG) pulse sequence[19], with an echo time of 10 ms and motion compensated in the slice direction, was used for both the flow and mixing measurements. For flow measurements, the sequence included a velocity encoding bipolar pair of PFG, with 1-ms-duration (δ) of the PFG pulse and 6.1 ms time separation (Δ) between the PFG pair. The PFG protocols used four PFG-pulse steps; the gradient strengths were 0, 4150, 8300 and 49840 Hz/cm, respectively.

In order to reduce the imaging time and the overall consumption of the fluid samples, no signal averages were taken for the MRI of this work. The acquisition time for a 2D image in the extruder (256 × 256 complex points) was 12.8 min, and 9 min, with a spatial resolution of 0.02 × 0.02 × 0.3 cm, for 20 (repetition time 2 s) and 30 rpm (repetition time 3 s), respectively. The total time for acquisition of a set of 2D maps of velocities in all x, y and z directions was 2 hour 12 minutes and 1 hour 30 minutes, respectively.

For MR imaging of mixing, no PFG gradients were applied and a pair of MR images under identical flow conditions for the major-component were acquired. The first image was with flow only of the major component whereas the second had flow of both components and hence of their mixing. The acquisition time for a 2D image (256 × 256, spatial resolution 0.02 × 0.02 × 0.3 cm, no signal average) was 9 min for rotation at 30 rpm and 12.8 min for 20 rpm.

2.4 Calibration of SI-C Relationship

Prior to the mixing MRI, the relationship between the MRI signal intensity (SI), the $MnCl_2$ concentration (C, mmol/l) and the receiver gain (RG) was calibrated by imaging a bundle of 8 glass vials each filled with 1% aqueous CMC containing different concentrations of $MnCl_2$ (0.1, 0.2, 0.4, 0.8, 1, 1.4, 1.6, 2 mmol/l) using the same protocol as the mixing MRI, which was repeated 6 times each with a different receiver gain value (17, 20, 23, 26, 28, 30). The relationship obtained by linear regression was:

$$\text{Log(SI)} = 5.38 - 0.05\ \text{RG} - 0.37\ \text{C}, \qquad R^2 = 0.99, \qquad (1)$$

In this work, all the MRI measurements used the same RG values, since this allowed the subsequent data processing and analysis to be much simpler.

Although Equation (1) can be used directly to convert image signal intensities into concentrations, inter-comparisons between those concentration maps obtained under different flow conditions required acquisition of a blank "flow only" image before that of the "mixing". In addition, this blank "flow only" image enabled the spatial segmentation of the extruder chambers from the background in the image, which is essential for the statistical analysis of the mixing index.

2.5 Formulation of the Mixing Indices

Mixing indices were calculated from $MnCl_2$ concentration maps, obtained from the mixing images and SI-C calibration (Equation 1), either using the expression[1,5]:

$$I_1 = S^2 / [C_{mean}(1 - C_{mean})], \qquad 0 \le I_1 \le 1; \qquad (2)$$

or using the expression[5]:

$$I_2 = S^2 / (p \times q), \qquad 0 \le I_2 \le 1; \qquad (3)$$

where S^2 is the variance of concentrations of the mixture and C_{mean} is the mean concentration estimated from sampling; p and q are the proportions of the two components estimated from samples. In both Equations (2) and (3), the indices I_1 and I_2 are equal to zero when the system is fully mixed and are equal to 1 when completely segregated.

For this work, both Equations (2) and (3) were used for comparison. In Equation (2), C_{mean} is defined as the flow-rate-weighted mean concentration[20]:

$$C_{mean} = \frac{Q_{ma}.C_{ma} + Q_{mi}.C_{mi}}{Q_{ma} + Q_{mi}}. \qquad (4)$$

For Equation (3), p is the flow-rate-weighted proportion of the major component[20]:

$$p = \frac{Q_{ma}}{Q_{ma} + Q_{mi}}; \qquad (5)$$

and q is that of the minor component (i.e. $q = 1 - p$); where Q_{ma} and Q_{mi} are the flow rates of the major and minor components respectively, and C_{ma} and C_{mi} are the $MnCl_2$ concentrations of the major and minor components respectively. S^2, in Equations (2) and (3), is the variance of concentrations of the mixture in a concentration map of the extruder, that is, the sum of variances in each part of the extruder shown in the image obtained using image analysis software developed in house.

3 RESULTS AND DISCUSSION

3.1 Flow Measurements

Transverse and coronal 2D MR images of flow of the 1% CMC solution (made by 0.1 mmol/l $MnCl_2$), through the region near either the inlet or the outlet of the screw extruder, were acquired at 21^0 C using the GE-PFG imaging sequence. The flow rate was 52 ml/min and rotation speed 30 rpm. The velocity components in the x, y and z directions (Vx, Vy and Vz) were measured and velocity maps were reconstructed. Figure 1 shows a representative set of velocity maps of a coronal slice near the inlet region with the grey scales representing the values of velocity (cm/s). These data were then used to construct plots in which the 2D velocity vector maps were overlaid on the speed contours using Matlab software (speed = $(v_x^2 + v_y^2 + v_z^2)^{1/2}$). Figure 2 shows a representative pair of plots from the region near the inlet of the extruder with the grey scales corresponding to the values of speed (cm/s); Figure 2A presents those from a transverse (xy) slice, and Figure 2B those from a coronal (zx) slice. Those same velocity maps were also converted into 3D velocity vectors using Matlab (not shown). Display of MRI velocity data as speed contours and velocity vectors facilitates easier comparison of MRI results with computational fluid dynamic (CFD) simulations, and hence provides means for validation of CFD software.

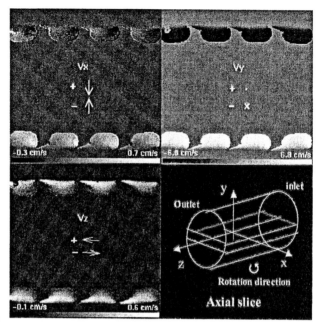

Figure 1 *Representative 2D velocity maps of Vx, Vy and Vz of an axial slice near the inlet region of the extruder with the flow rate 52 ml/min and rotation speed 30 rpm.*

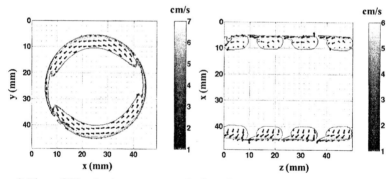

Figure 2 *Plots of 2D velocity vectors overlaid on the speed contours constructed from MRI velocity maps: A. those from a transverse (xy) slice, and B. those from a coronal (zx) slice near the inlet region of the extruder with the flow rate 52 ml/min and rotation speed 30 rpm (anticlockwise).*

3.2 MRI of Mixing

Coronal 2D MRI slices near either the outlet or the inlet region of the screw extruder were acquired during mixing in the extruder using a range of flow rates and screw rotation speeds. For each set of mixing conditions, a pair of "flow only" and "mixing" images was acquired with identical receiver gain settings. Maps of $MnCl_2$ concentration reconstructed from that pair of images using Equation (1) are shown in Figures 3 and 4 in which the grey scale shows the concentration of $MnCl_2$ (mmol/litre), with the highest concentration (1.6

mmol/l) the brightest, and the lowest concentration darkest. The images in Figure 3 were acquired near the inlet, whereas those in Figure 4 were near the outlet. Figures 3A and 4A are for 30 rpm rotation, 4.5 ml/min minor flow rate and 47 ml/min major flow rate; in contrast, the conditions for Figures 3B and 4B are 20 rpm, 7.4 ml/min minor flow rate and 76 ml/min major flow rate.

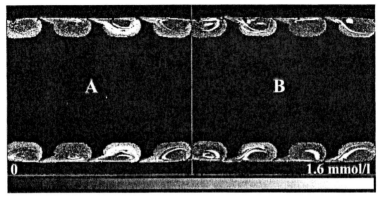

Figure 3 *Calculated concentration maps near the inlet: A. obtained with 30 rpm rotation, 4.5 ml/min minor flow rate and 47 ml/min major flow rate; and B. 20 rpm, 7.4 ml/min minor flow rate and 76 ml/min major flow rate.*

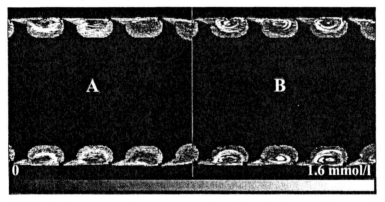

Figure 4 *Calculated concentration maps near the outlet: A. the same flow conditions as Fig. 3A; and B. the same flow conditions as Fig. 3B.*

Table 1 *MRI measured mixing parameters*

Position	Rotation speed (rpm)	Total flow rate (ml/min)	C_{mean} (mmol/l)	C_{max} (mmol/l)	I_1	I_2
Near outlet	20	83	0.20	1.58	0.62	0.47
	30	52	0.23	1.27	0.34	0.25
Near inlet	20	83	0.18	1.60	0.73	0.55
	30	52	0.22	1.25	0.47	0.35

Under above flow conditions, the mean concentration (C_{mean}) when fully mixed was 0.23 mmol/l. From the MRI concentration maps (Figures 3 and 4), the indices of mixing (I_1 in Equation (2) and I_2 in Equation (3)) were calculated and are listed in Table 1, together with the maximum and mean concentrations calculated from those maps. The values of the mixing indices were in the range from 0.34 to 0.73 (I_1) or 0.25 to 0.55 (I_2) under different flow conditions.

It is clear from Equations (2) and (3) that the smaller the value of the index the better is the quality of the mixing. Table 1 shows that the smallest value of mixing index (either I_1 or I_2) corresponds to the faster speed (30 rpm) and slower flow rate (52 ml/min) near the outlet; in contrast, the largest index value is from near the inlet with the slower speed (20 rpm) and faster flow rate (83 ml/min). These measurements are consistent with expectation; the fact that the same relative order comes from the calculations using either Equation (2) or Equation (3) shows that both equations give the same estimation of mixing quality, even though the index values are numerically different.

4 CONCLUSIONS

This study clearly demonstrates that MRI is a powerful tool for non-destructive, quantitative measurements of flow velocities and the quality of mixing in a single-screw extruder. MRI velocity data displayed as speed contours and velocity vectors provides easier comparison of MRI results with CFD simulations and hence are a valuable mean for validation of CFD software.

The mixing index chosen for this work can be used for mixing of both miscible and immiscible fluids; it is encouraging that the experimental results are consistent with the expectation. Furthermore, this MRI method will not only aid future studies of mixing in screw extruders, but also in many other types of configurations used for mixing.

Acknowledgements
The authors thank the late Dr. Herchel Smith for a munificent benefaction.

References

1   M. Tyszka, R. C. Hawkes and L. D. Hall, *Fluid Meas. and Instrum.*, 1991, **2**, 131.
2   E. Fukushima, *Annu. Rev. Fluid Mech.*, 1999, **31**, 95.
3   P. T. Callaghan, *Rep. Prog. Phys.*, 1999, **62**, 599.
4   L. F. Gladden, *Topics in Catalysis*, 2003, **24**, 19.
5   C. K. Agemura, R. J. Kauten and K. L. McCarthy, *J. Food Eng.*, 1995, **25**, 55.
6   Corbett A M, Phillips R J, Kauten R J and McCarthy, K L *J. Rheol.*, 1995, **39**, 907.
7   K. Rombach, S. Laukemper-Ostendorf and P. Blümler, 1998 in: *Spatially Resolved Magnetic Resonance Methods,Materials, Biology, Rheology, Geology, Ecology, Hardware*, eds. P. Blümler, B. Blümich R. Botto and E. Fukushima, Willy-VCH, 1998, 517.
8   M. H. G. Amin, A. D. Hanlon, L. D. Hall, C. Marriott, S. Ablett, W. Wang and W. J. Frith, *Meas. Sci. Technol.*, 2003, **14**, 1760.
9   E. G. Smith, R. Kohli, P. A. Martin, T. Instone, N. Roberts and R. H. T. Edwards, in: *Frontiers in Industrial Process Tomography*, eds. D. M. Scott and R. A. Williams, Engineering Foundation, 1995, p.223.

10 W. Wang, J. H. Walton, and K. L. McCarthy, *J. Food Proc. Eng.*, 2000, **23**, 403.

11 S. Middleman, *Fundamentals of Polymer Processing,* New McGraw-Hill Inc, 1977.

12 C. Rauwendaal, *Polymer Extrusion*, Carl Hanser Verlag, 1986.

13 J. M. Harper, *Extrusion of Foods, Volume 1,* CRC Press Inc, 1981.

14 M. N. Riaz, *Extruders in Food Applications,* Technomic Publishing Company, 2000.

15 N. Harnby, M. F. Edwards, and A. W. Nienow, *Mixing in the Process Industries*, Butterworth-Heinemann, 1992.

16 R. J. McDonough, *Mixing for the Process Industries,* Van Nostrand Reinhold, 1992.

17 Z. Tadmor and C. G. Gogos, *Principles of Polymer Processing,* John Wiley & Sons Inc, 1979.

18 R. W. Whorlow, *Rheological Techniques*, John Wiley & Sons, 1980.

19 P. T. Callaghan, *Principles of Nuclear Magnetic Resonance Microscopy*, Clarendon Press, 1991.

20 M. H. G. Amin, L. D. Hall, W. Wang and S. Ablett, *Meas. Sci. Technol.,* 2004, **15,** (in press).

ESR AND NMR SPECTROSCOPY STUDIES ON EMULSIONS CONTAINING β-LACTOGLOBULIN AND OXIDISED METHYL LINOLEATE

Nazlin K. Howell, Suhur Saeed and Duncan Gillies

University of Surrey, School of Biomedical and Molecular Sciences, Guildford, Surrey GU2 7XH, United Kingdom. E-mail: N.Howell@surrey.ac.uk

1 INTRODUCTION

The properties of gels and emulsions are governed by food components present and their interactions with each other. In this study we investigated the interactions of a food whey protein, β-lactoglobulin with a polyunsaturated fatty acid methyl linoleate; this is important as manufacturers seek to replace saturated fats with unsaturated ones in dairy, meat and bakery products. A high fat intake is associated with atherosclerosis, colon and breast cancer whereas polyunsaturated fatty acids (PUFA) such as fish oils are reported to be beneficial in reducing the risk of coronary heart disease.[1]

However, a major problem associated with long chain polyunsaturated fatty acids is oxidation, which causes rancidity and toxic products.[2] Lipid oxidation occurs in the presence of oxygen, transition metals and enzymes.[3] In the initiation stage, a hydrogen atom is abstracted from a methylene group in the PUFA by a reactive species such as an hydroxy radical (OH˙) leaving behind an unpaired electron on the carbon (-CH- or lipid radical). The carbon radical is stabilised by a molecular rearrangement to form a conjugated diene which can combine with oxygen to form a peroxy radical ROO˙ that can abstract further H from other lipid molecules to form lipid hydroperoxides and cyclic peroxides. Hydroperoxides are unstable and degrade to secondary products including hydroxy fatty acids, epoxides and scission products such as aldehydes (malondialdehyde), ketones and lactones, many of which are toxic.

Metal ions can initiate the degradation of hydroperoxides and can also generate further free radicals via the Fenton reaction. Lipid radicals can combine with each other or more likely with protein molecules to end the chain reaction. The latter reaction causes cross-linking and severe damage to proteins. The free radical chain reaction can also be terminated by the addition of antioxidants.

Antioxidants inhibit lipid oxidation through competitive binding of active forms of oxygen involved in the initiation step of oxidation. Alternatively, antioxidants can react with fatty acid peroxyl free radicals (LOO˙) to form stable antioxidant-radicals (TO˙) which are either too unreactive for further reactions or form non-radical products. Propagation steps may be retarded by destroying or inhibiting catalysts, or by stabilising hydroperoxides.[3,4]

Protein-lipid interactions

Early studies indicated that lipid oxidation products including hydroperoxides, aldehydes, ketones and epoxides react with amines, thiols, disulphides and phenolic groups in the protein via covalent bonds.[5] Non-covalent hydrogen bonds involving hydroxyl and carboxyl groups in proteins and lipids and a loss of specific amino acids such as cysteine, lysine, histidine and methionine as well as cross-linking and DNA damage are also reported. Subsequent studies confirm that protein oxidation is prolific in oxidising systems e.g. containing $FeCl_3/H_2O_2/$ ascorbate.

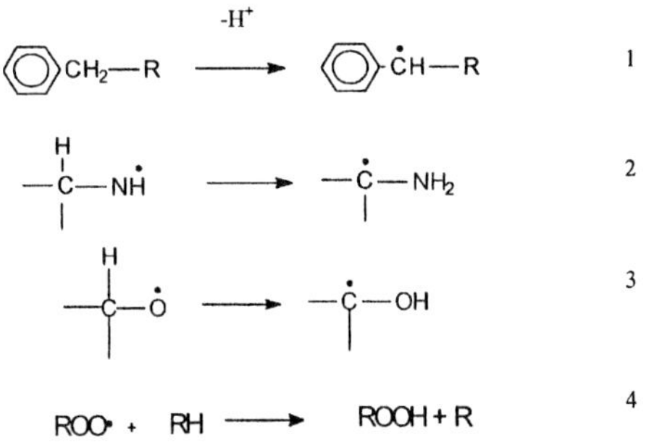

$$\underset{}{\bigcirc\!\!\!-CH_2-R} \xrightarrow{-H^+} \underset{}{\bigcirc\!\!\!-\overset{\bullet}{C}H-R} \qquad 1$$

$$\underset{H}{\overset{|}{-C}-N\overset{\bullet}{H}} \longrightarrow -\overset{\bullet}{\underset{|}{C}}-NH_2 \qquad 2$$

$$\underset{H}{\overset{|}{-C}-\overset{\bullet}{O}} \longrightarrow -\overset{\bullet}{\underset{|}{C}}-OH \qquad 3$$

$$ROO^{\bullet} + RH \longrightarrow ROOH + R \qquad 4$$

Scheme 1 *The formation of carbon-centred radical via hydrogen abstraction (1) or via secondary reactions of other species, such as further reaction of nitrogen centred radical (2), alkoxy (3) and peroxyl radical (4)*

Our previous studies in model systems indicate that a wide range of amino acids are capable of forming free radicals in the presence of oxidising methyl linoleate and fish oils with subsequent cross-linking as ascertained by fluorescence spectroscopy.[6] However, aromatic amino acid residues are particularly sensitive to oxidation. The adduct species formed within the aromatic rings are stabilised by delocalisation onto neighbouring double bonds.[7] In the absence of sufficient oxygen, the major fate of carbon-centred radicals formed on proteins is dimerisation. The amino acid most likely to be damaged is tyrosine resulting in cross-linking to form dityrosine. Tyrosine phenoxyl radicals are highly stabilised/delocalised radicals and result in dimerisation through the carbon atom on the aromatic ring.

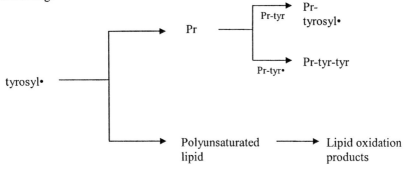

Scheme 2 *Role of tyrosyl free radicals in the formation of protein (Pr) cross-links, dityrosine and lipid oxidation products (adapted from Ref.12)*

Dityrosine is a protein oxidation marker and is associated with a wide range of disorders and ageing, suggesting that tyrosyl radicals plays an important role in initiating and propagating oxidation of proteins. Most oxidising agents, including ultraviolet irradiation, promote the *in vitro* and *in vivo* formation of dityrosine. The tyrosine radicals formed can go on to attack further amino acids and lipids resulting in a autocatalytic chain reaction (Scheme 2).[7]

Badii and Howell[8] confirmed using antioxidant treatments that fish toughening during frozen storage may be attributed to the protein-oxidised lipid interactions. Howell and co-workers are currently examining the effects of protein-oxidised lipid interactions on texture and emulsification in other proteins.

The objectives of the present study were to investigate the nature of free radicals produced in β-lactoglobulin (β-LG) protein-oxidised lipid methyl linoleate (ML) mixtures, with and without antioxidants by ESR spectroscopy; to study the effect of free radical damage of β-lactoglobulin by oxidised lipids by 1D and 2D NMR spectroscopy and to confirm changes in β-lactoglobulin by FT-Raman spectroscopy. The following results are summarised from a paper submitted for publication by Saeed *et al.*[9]

2 MATERIALS AND METHODS

2.1 Sample preparation

β-LG (10-15% w/w) in distilled water was mixed with fresh ML (10%) to form an emulsion using the Sorvall Omni mix homogeniser speed 4 for 5 min. A separate experiment was performed as above, containing ML oxidised under UV light for 24, 48 and 72 h. The samples were tested at time 0 and after storage for different lengths of time, depending on the techniques used below. The effect of antioxidants was tested by the addition of 250 ppm each of vitamin C, vitamin E and a mixture of vitamins C + E.

2.2 ESR spectral measurements

The formation of spin trap radical adducts was monitored by ESR spectroscopy. Spin trap 2-methyl-2-nitrosopropane (MNP) in 25mM dimethylsulfoxide (DMSO) was added to the emulsion. The mixtures were incubated for 30 min at 22 °C. ESR spectra were recorded on a Jeol RE IX X-band spectrometer, operating at 9.71 GHz with 100 kHz magnetic field modulation.

2.3 NMR spectroscopy

Conformational changes occurring in β-lactoglobulin in the absence and presence of ML were studied by 1D and 2D ^{1}H *NMR spectra* (Double quantum filtered-COSY (DQF-COSY)) using a Bruker 500 MHz NMR spectrometer.

2.4 FT-Raman spectroscopy

Emulsions prepared as above were examined on a Perkin-Elmer System 2000 FT-Raman Spectrophotometer with excitation from Nd:YAG laser at 1064 nm and laser power of

1785 mW. Frequency calibration of the instrument was performed using the sulphur line at 217 cm^{-1}.

3   RESULTS AND DISCUSSION

3.1 ESR spectroscopy

In the present study, a nitroxide radical was detected when MNP was added to the reaction mixture of β-lactoglobulin and 24h oxidised methyl linoleate (ML) (Figure 1). An ESR signal was not detected for protein alone, in the absence of oxidised ML, indicating that the detected nitroxide radical depended on the hydroxyl radical produced from oxidised ML.[9] Only a 3–line ESR spectrum was observed and additional hyperfine coupling was not seen, showing that the radical was centred on a tertiary carbon, in particular tyrosine. Within 2 h the transfer of a radical adduct to a second species, presumably protein was evident (Figure 1a). After 24 h the radical reached its maximum intensity (Figure1b) and remained stable when stored at -20°C, but decreased considerably when stored at 22 °C (Figure 1c). Antioxidants prevented the transfer of the radical to the protein as shown by the shape and reduced size of the radical adduct (Figure 1 d,e).

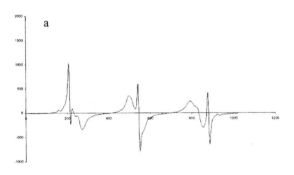

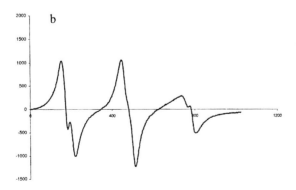

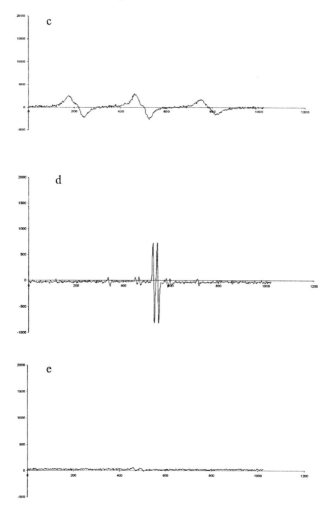

Figure 1   *ESR spectra of mixtures of β-lactoglobulin (15% w/win distilled water) and 24 h oxidised methyl linoleate (10%), obtained after the addition of MNP a) 2h later, b) 24 h later, c) after one week at 22 °C (d) with vitamin C and (e) with vitamins C + E. Typical instrument settings were as follows, if not stated otherwise. Microwave power (4 mW); modulation amplitude (0.5 mT); time constant (0.01s); gain (200); sweep width, (10 mT); 20 scans.*

Antioxidant effects varied; vitamin C had the biggest effect on radical reduction whereas vitamin E had the least effect. Synergistic interaction was observed with a mixture of vitamins C + E resulting in a major reduction in the free radical signal. Vitamin E compounds (tocopherols and tocotrienols) are reported to effectively inhibit lipid oxidation in foods and biological systems. In addition, the protective role of ascorbic acid in biological systems via free radical reactions has been studied.[3] In many systems including lipid oxidation, Vitamins E and C act synergistically.[8] Vitamin E is lipophilic and is considered to be the primary antioxidant. Vitamin C reacts with the vitamin E radical to regenerate vitamin E, and the resulting ascorbic acid radical is reduced back to vitamin C by NADH. The chemistry and antioxidant properties of tocopherol and ascorbic acid have been summarised by Howell and Saeed.[4]

3.2 NMR spectroscopy

Interesting changes were observed within the amide and aromatic regions, using controlled exchange of protons for deuterons from D2O solvent. A proton can only exchange if it is accessible to the solvent and H-bonded protons only exchange if there is an unfolding transition that allows them to contact the solvent. NMR spectra indicated deuterium exchange and NH resonance belonging to several regions of β-LG (Figure 2 a,b). Oxidation and protein denaturation resulted in an increase in structural flexibility and some initially protected backbone amide groups were exposed thus becoming sharper and easily identifiable.[9]

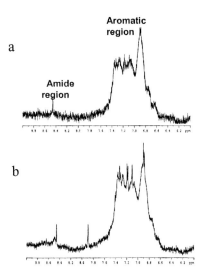

Figure 2   *NMR spectra of a) β-lactoglobulin-ML emulsion b) β-lactoglobulin-oxidised ML emulsion*

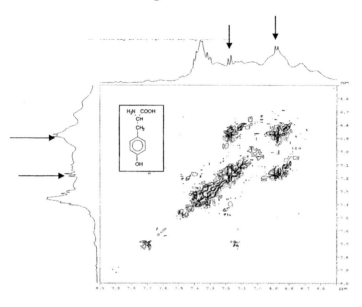

Figure 3   *DQF-COSY spectra of the aromatic region of β-lactoglobulin-ML emulsion. Two sets of protons from tyrosine are highlighted (adapted from Ref.9).*

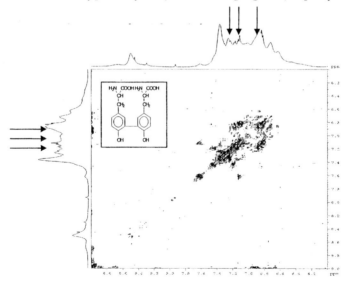

Figure 4   *DQF-COSY spectrum of the aromatic region of β-lactoglobulin-oxidised ML emulsion. Three different sets of protons agree with the structure of dityrosine (adapted from Ref. 9).*

DQF-COSY spectra indicated the dimerisation of tyrosine in the unoxidised sample (Figure 3) to dityrosine in the oxidised emulsion (Figure 4).[9] Dityrosine is an oxidation marker which is associated with many diseases including heart disease and diabetes as well as ageing; dityrosine production in the presence of oxidised lipids has been reported for the first time by Saeed *et al.*[9]

3.3 Raman spectroscopy

Raman spectra (1800-600 cm^{-1}) of protein-oxidised lipid mixtures, when compared to protein-fresh methyl linoleate samples exhibited changes in the amide region, showing a reduction in α-helix and an increase in β-sheet; this indicated denaturation of the protein molecule.[10] The unfolding of the protein exposed the buried tryptophan residues.[11] The tyrosine doublet ratio I $_{850/830}$ was lower in the oxidised sample compared with the unoxidised protein suggesting that the phenolic hydroxyl oxygen is ionised or strongly hydrogen bonded to a negatively charged molecule.[9-11] Raman spectra confirmed changes in the amide and aromatic groups of the protein, reported by NMR and ESR spectroscopy above.

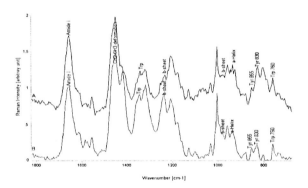

Figure 5 *Raman spectra of a mixture of β-lactoglobulin (15%) and A) fresh methyl linoleate (10%) and B) oxidised methyl linoleate*

4 CONCLUSIONS

ESR spectroscopy is a useful tool for the characterisation of different free radicals from oxidised lipid, proteins and antioxidant, and for showing the transfer of free radicals from the lipid to the protein during storage, and reduction of the free radicals by antioxidants. NMR and FT-Raman spectroscopy confirmed major changes in protein secondary structure and hydrophobic groups in the presence of oxidised lipids. Further detailed analysis is underway to identify the extent of damage to proteins and amino acids due to oxidised lipids.

Acknowledgements

This research project was funded by a BBSRC grant awarded to Prof. Nazlin K.Howell. The authors thank Mr. Jim Bloxsidge for technical assistance with the NMR analysis as well as Dr G. Wagner, University of Surrey, School of Biomedical and Molecular Sciences and Prof. P. Belton, University of East Anglia for useful discussions on the NMR data.

References

1.   R. Doll, and J. Peto, *J. National Cancer Institute*, 1981, **66**, 1192.
2.   M.G. Simic, S.V. Jovanovic and E. Niki, Mechanisms of lipid oxidation processes and their inhibition. In *Lipid oxidation in food*, ed., St. Angelo, A. J., ACS Symposium Ser. 500, Am.Chem. Soc. Washington, DC 1992 pp 55.
3.   B. Halliwell, and J..M. C.Gutteridge, Free radicals in Biology and Medicine. 3rd edn. Clarendon Press, Oxford, 1999.
4.   N.K. Howell and S. Saeed, The effect of antioxidants. In: *Antioxidants in Human Health and Disease*. Ed. T.K. Basu, N.J. Temple and M.L. Garg. CAB International, Oxford, UK. 1999 pp 43-54.
5.   H.W. Gardner, Lipid hydroperoxide reactions - a review. *J. Agric. Food Chem.,* 1979, **27**, 220-229.
6.   S. Saeed, S. Fawthrop and N.K. Howell, ESR study on free-radical transfer. J. Sci. Food Agric. 1999, **79**, 1809-1816.
7.   M. Sharma and R. Jain, Isolation and analysis of dityrosine from enzyme-catalysed oxidation of tyrosine and X-irradiated peptide and proteins. Chemico-Biological Interactions, 1998, **108**, 171-185.
8.   F. Badii, and N. K.Howell, Effect of antioxidants, citrate and cryoprotectants on protein denaturation and texture of frozen cod (*Gadus morhua*). J.Agric. Food Chem., 2002, **50**, 2053-2061.
9.   S. Saeed, D. Gillies, and N.K. Howell, ESR and NMR spectroscopy studies on the formation dityrosine in emulsions containing proteins and oxidised methyl linoleate. Arch. Biochem. Biophys., (2004). Submitted.
10. N.K. Howell and E.C.Y. Li-Chan, Elucidation of the interaction of lysozyme and whey proteins by Raman spectroscopy. Int. J. Food Sci. Technol., 1996. 31 439-452.
11. N.K. Howell, H.Herman, and E.C.Y. Li-Chan, Elucidation of protein-lipid interactions in a lysozyme-corn oil system by FT- Raman spectroscopy. J. Agric. Food Chem., 2001 49, 1529-1533.
12. J.W. Heinecke, Tyrosyl radical production by myeloperoxidase: a phagocyte pathway for lipid peroxidation and dityrosine cross-linking of proteins. Toxicology, 2002, **177**, 11-22.

PARAMAGNETIC CHALLENGES IN NMR MEASUREMENTS OF FOODS

Stefano Alessandri[1,2], Francesco Capozzi[1,3], Mauro A. Cremonini[3], Claudio Luchinat[1,2], Giuseppe Placucci[3], Francesco Savorani[3], Maria Turano[3,4].

[1]Magnetic Resonance Centre CERM, University of Florence, Italy
[2]Department of Agricultural Biotechnology, University of Florence, Italy
[3]Department of Food Science, University of Bologna, Italy
[4]Department of Chemistry, University of Calabria, Italy

1 INTRODUCTION

NMR spectroscopy is more and more used in the evaluation of food quality, although all its potentialities have not yet been exploited. Paramagnetic metal ions, either naturally present or added to foodstuff, provide an additional source of information that can help to characterize the food.

The presence of "paramagnetic" signals in spectra of complicated systems is extremely helpful for their interpretation and provides information otherwise not obtainable. In complex mixtures, NMR spectra are diagnostic, for example, for the oxidation and spin states of iron porphyrins in myoglobin and hemoglobin, yielding information about the metal ion coordination sphere.[1] In turn, the latter determines the extent to which paramagnetism affects the relaxation rates of those water molecules that are able to bind protein sites in proximity of the heme pocket; the relaxation enhancement is then transmitted to the bulk water molecule by chemical exchange.[2]

Paramagnetic metal ions, like some lanthanides, can also be added to samples submitted to NMR investigation in order to simplify the spectra, and to easily identify species which are present in the sample but hidden from the overlapping signals of the same large molecule or belonging to other more abundant substances present in a mixture. For instance, Yb^{3+} replaces calcium ions in many biomolecules and its effects have been exploited to obtain distance information, increasing the quality of NMR structures.[3]

Nuclear relaxation is deeply affected by coupling with unpaired electrons: the observed relaxation rates of water protons present in food could be highly increased by the presence of metal ions, provided that water molecules have access to them (inner solvation sphere) or to their proximity (outer solvation sphere). Keeping this in mind, all explanations for different relaxation times measured on food with LF-NMR should take into account possible paramagnetic effects, especially for those cases in which the concentration of the metal, as well as its spin state and water accessibility, are expected to give a contribution comparable with the diamagnetic one.

Some possible exploitations of the paramagnetic effect in food science will be suggested, hoping that their use, already successful in structural biology, will soon bring to meaningful results in the food area.

2 PARAMAGNETIC COMPOUNDS

Paramagnetic compounds are characterized by the presence of unpaired electrons, usually belonging to one ore more transition metal ions found ubiquitariously in many biological systems such as those constituting the prosthetic group of proteins. Radicals are also paramagnetic but they are not suitable for high-resolution NMR experiments, the most important technique used to investigate in detail the structure of macromolecules in solution. Nowadays, the presence of paramagnetic metal ions coordinated by macromolecules (i.e. metallo-proteins) not only does not constitute a problem anymore (loss of information around the metal due to large broadening of signals and poor NOE effect due to short longitudinal relaxation times) but often it is looked for (also by metal substitution) since it provides additional constraints for structure elucidation procedures[4,5] Further information is accessible also with low-resolution experiments investigating water molecules interacting with the metal ion in rapid exchange with the bulk water: many features about the active site of metallo-enzymes are deduced by exploring the parameters obtained by relaxometry.

Although NMR spectroscopy is more and more used in food science, the paramagnetic effect is still far from being fully exploited, even though foodstuffs are unequivocally biological systems (no food no life!) and metal ions are necessary to life. It is not a matter of total amount (other elements are orders of magnitude more abundant) but they are essential to life because used as important cofactors for the functionality of enzymes. In Table 1, the abundance of paramagnetic metals in the human body is reported, together with important magnetic properties which will be discussed later.

Table 1   *Abundance of paramagnetic metals in human body[6]*

Element	mg/Kg	Paramagnetic Oxidation States [a]	Total Spin Quantum Number (S)
iron	60.0	Fe^{3+}(H.S.); Fe^{3+}(L.S.); Fe^{2+}(H.S.)	5/2; 1/2; 2
copper	1.00	Cu^{2+}	1/2
cerium	0.57	Ce^{3+}	1/2 (J=5/2)
titanium	0.29	Ti^{3+}	1/2
nickel	0.21	Ni^{2+}	1
chromium	0.20	Cr^{3+}; Cr^{2+}	3/2; 2
manganese	0.17	Mn^{3+}; Mn^{2+}	2; 5/2
cobalt	4.3 e -02	Co^{2+}(H.S.); Co^{2+}(L.S.)	3/2; 1/2
vanadium	1.6 e -03	VO^{2+}; V^{3+}; V^{2+}	1/2; 1; 3/2

[a]H.S.: high spin ground state; L.S.: low spin ground state

Iron is the most abundant transition metal and represents the first choice for many biological functions: it participates to electron transport chains and works also as an oxygen carrier in animal bodies. For this reason it is abundant, as heme iron in myoglobin and hemoglobin, in muscles and, later, in meat. Non-heme iron is found in spinach as ferredoxin and in egg albumen bound to transferrin. Table 2 lists the content of three important paramagnetic metals, at least in one oxidation state, in some representative foods.

Not only metals are nutritionally important but they can also determine the quality of food by altering the shelf life (e.g. by catalyzing the oxidation of lipids[8]) or by giving unwanted colorations during storage.[9]

The colour of meat is largely determined at the meat surface by the relative amounts of three forms of myoglobin, i.e. deoxymyoglobin (DeoxyMb), metmyoglobin (MetMb), and oxymyoglobin (OxyMb). During external processes such as cooking, storage, and irradiation, the three forms of myoglobin interconvert and are degraded through oxygenation, oxidation and reduction reactions, ultimately influencing the appearance of meat colour. Livingston and Brown observed that increased OxyMb was stable under high oxygen conditions but the presence of DeoxyMb greatly increased the susceptibility of heme to auto oxidation to MetMb[10]. The three forms give different ^{1}H-NMR spectra because the iron ions have different oxidation/spin states. Only the OxyMb is diamagnetic (low spin iron(II)), whereas the other two forms have unpaired electrons on the iron ion: high spin iron(II) (DeoxyMb) and high spin iron(III) (MetMb). The latter gives an adduct with cyanide which possesses low spin iron(III) and is characterized by an ^{1}H-NMR spectrum with the sharpest paramagnetic signals.

Table 2 *Main paramagnetic metals in representative foods*[7]

Foods	Iron (mg/Kg)	Copper (mg/Kg)	Manganese (mg/Kg)
Beef loin	[a]77.0 (47.0)	1.10	0.11
Spinach	33.4	0.77	5.40
Bread, whole wheat	33.1	2.53	21.9
Eggs	21.3	0.66	0.31
Pork loin	[a]19.0 (11.0)	0.91	0.11
Chicken meat (breast)	[a]15.0 (6.00)	0.54	0.20
Potato with peel	11.7	0.88	2.27
Asparagus	5.90	1.13	1.45
Potato without peel	3.30	0.54	1.18
Peach fruit	2.50	0.67	0.58
Orange fruit	1.50	0.43	0.29
Orange juice	1.00	0.34	0.34
Raw milk	0.10	0.05	0.02
Beer	0.00	0.05	0.10
Tap water	0.00	0.02	0.00

[a]Total iron (heme iron)

Myoglobin will be used to describe the effect of the oxidation states and the coordination chemistry on the ^{1}H-NMR spectrum.

3 THE ORIGIN OF THE PARAMAGNETIC EFFECT: THE ELECTRON-NUCLEUS COUPLING

3.1 The isotropic shift

The chemical shifts of nuclei belonging to ligands bound to paramagnetic ions may be very far away (even thousands of ppm!) with respect to those measured on their corresponding diamagnetic analogues, depending on the type of metal ion and its spin state. This is due to the contributions of the Fermi *contact* term to the *isotropic shift*.[11,12] The latter extends also to nuclei of molecules not directly bound to the paramagnetic ion but sensing its unpaired electron magnetic moment through space, with an effect, the dipolar coupling (whose

averaged effect in solution is called *pseudocontact* shift), decreasing with the third power
of the distance between the metal ion and the observed nucleus.

The isotropic shift is experimentally determined by subtracting, for each nucleus, the
shift measured in a "diamagnetic reference" from that measured in the paramagnetic
protein. The "diamagnetic reference" is usually the diamagnetic form of the same protein,
in which either a diamagnetic metal substitutes for the paramagnetic one in one or more
binding site (as in Ca binding proteins)[3] or where a metal has been made diamagnetic by
changing its oxidation state (as in cytochromes)[1].

The Fermi contact shift depends on the orientation of the molecule in the magnetic field if
g is anisotropic and different from g_e, the Lande factor, which expresses the proportionality
between angular and magnetic moments of electron. In the case of a single S manifold,
without zero field splitting, the contact shift is given by:

$$\delta^{con} = \frac{A}{\hbar} \frac{g_e \mu_B S(S+1)}{3\gamma_I kT} \qquad (1)$$

where S is the total spin quantum number (see Table 1 for values associated to some
important metal ions), k is the Boltzmann's constant, T the absolute temperature, μ_B, the
Bohr magneton, which represents the conventional unit of measurement of magnetic
moments on a microscopic scale, equal to 9.2741 x 10^{-24} JT^{-1}, $\hbar$ is the Planck's constant,
and γ_I is the nuclear magnetogyric ratio.

The contact coupling constant, A, is related to the total unpaired spin density on a nucleus,
ρ, through the following equation:

$$A = \frac{\mu_0}{3S} \hbar \gamma_I g_e \mu_B \rho \qquad (2)$$

where μ_0 is the magnetic permeability of a vacuum, and the other symbols have been
already defined.

The pseudocontact shifts of nuclei in paramagnetic proteins depend on the anisotropy
of the magnetic susceptibility tensor (χ).The relationship between pseudocontact shifts
(δ_{pc}) and the polar coordinates of a nucleus in the principal axes system of the
susceptibility tensor is:

$$\delta_{pc}^i = \sum_j \frac{1}{12\pi \cdot r_{ij}^3} \left[\Delta\chi_{ax}^j (3n_{ij}^2 - 1) + \frac{3}{2} \Delta\chi_{rh}^j (l_{ij}^2 - m_{ij}^2) \right] \qquad (3)$$

where $\Delta\chi_{ax}$ e $\Delta\chi_{rh}$ are the axial and rhombic components of the susceptibility tensor
anisotropies, l_{ij}, m_{ij} and n_{ij} are the direction cosines of the position vector of atom i with
respect to the jth magnetic susceptibility tensor coordinate system and r_{ij} is the distance
between the jth paramagnetic centre and nucleus i. The above equation is exploited as a
tool for structural characterization of metal binding proteins in the immediate vicinity of
the paramagnetic centre but not directly bound to it.

3.2 Nuclear relaxation rate enhancements

Nuclear relaxation rate enhancements occur whenever there are unpaired electrons.
Unpaired electrons generate magnetic fields that are sensed by nuclei through contact and
dipolar mechanisms. Fluctuations of these fields in time, with time constant τ_c, cause

nuclear relaxation. The rate enhancements for the longitudinal (R_1) and the transverse (R_2) relaxations due to contact mechanisms are, with some approximations:[13]

$$R_{1M}^{con} = \frac{2}{3}S(S+1)\left(\frac{A}{\hbar}\right)^2 \frac{\tau_c}{1+\omega_S^2\tau_c^2} \quad (4)$$

$$R_{2M}^{con} = \frac{1}{3}S(S+1)\left(\frac{A}{\hbar}\right)^2 \left[\frac{\tau_c}{1+\omega_S^2\tau_c^2} + \tau_c\right] \quad (5)$$

where ω_S is the electron Larmor frequency and the other symbols have been already defined.

The correlation time τ_c reflects the magnetic field fluctuation induced at the nucleus by the electron magnetic moment: changes of the electron magnetic moment orientation or interruptions of magnetic coupling due to chemical exchange, if present:

$$\tau_{c\ con}^{-1} = \tau_s^{-1} + \tau_M^{-1} \quad (6)$$

where τ_s is the electronic correlation time and τ_M is the time constant for chemical exchange. In the point dipole metal centred approximation, the relaxation enhancements are[15,16]:

$$R_{1M}^{dip} = \frac{2}{15}\left(\frac{\mu_0}{4\pi}\right)^2 \frac{\gamma_I^2 g_e^2 \mu_B^2 S(S+1)}{r^6}\left[\frac{7\tau_c}{1+\omega_S^2\tau_c^2} + \frac{3\tau_c}{1+\omega_I^2\tau_c^2}\right] \quad (7)$$

$$R_{2M}^{dip} = \frac{1}{15}\left(\frac{\mu_0}{4\pi}\right)^2 \frac{\gamma_I^2 g_e^2 \mu_B^2 S(S+1)}{r^6}\left[\frac{13\tau_c}{1+\omega_S^2\tau_c^2} + \frac{3\tau_c}{1+\omega_I^2\tau_c^2} + 4\tau_c\right] \quad (8)$$

Here τ_c is given by:

$$\tau_{c\ dip}^{-1} = \tau_S^{-1} + \tau_M^{-1} + \tau_r^{-1} \quad (9)$$

Rotation (τ_r), besides electron relaxation (τ_s) and chemical exchange (τ_M), is also a mechanism that modulates the electron magnetic field at the nucleus, because the electron and nuclear spins do not coincide in space.

In proteins, the correlation time for nuclear relaxation is almost always determined by τ_s since rotation is slow and exchange, if present, is also slower than τ_s. In order to obtain high-resolution spectra, the approximate borderline for τ_s is 10^{-11} s, even shorter for the largest S values. In the case of lanthanides the total - orbital plus spin - magnetic moment, J, should substitute S.

3.3 Solvent relaxometry

Several paramagnetic metalloproteins have one or more water molecules coordinated to the metal ion which are in exchange with the bulk solvent. The water proton longitudinal relaxation is given by:

$$R_{1obs} = R_{1dia} + R_{1para} \quad (10)$$

where R_{1obs} is the experimental value, R_{1dia} is the diamagnetic contribution and R_{1para} is the paramagnetic contribution. R_{1dia} can be determined using the apoprotein or a diamagnetic metal, *e.g.* Zn^{2+}. In turn,

$$R_{1para} = f\left(T_{1M} + \tau_M\right)^{-1} \qquad (11)$$

where f is the molar fraction of solvent bound to the metal, τ_M is the exchange time and $T_{1M} = R_{1M}^{-1}$ is the relaxation time of the protons of the solvent molecules bound to the metal ion.

4 ^{1}H-NMR SPECTRA OF MYOGLOBINS

The prosthetic group of myoglobins (Figure 1A) is the macrocycle protoheme: the iron atom is coordinated to this relatively rigid tetradentate macrocycle and easily accessible to monodentate ligands in both axial coordination positions (Figure 1B). The most common oxidation states available to heme iron are +2 and +3, each one existing in several spin states within the S=0 and S=5/2 range.

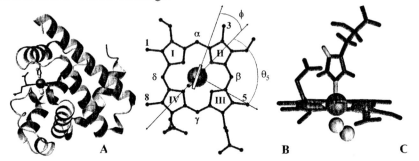

| **Figure 1** | *A: OxyMyoglobin; B) the macrocycle protoheme and the orientation of the proximal histidine; C) Hexacoordinated iron in OxyMyoglobin* |

4.1 High-Spin iron(III) in MetMb

In this form the heme iron is pentacoordinated with the His ligand in axial position or hexacoordinated with a water molecule in the sixth coordination site. The magnetic anisotropy is negligible and the observed hyperfine shifts are due primarily to the contact contribution.[17] Therefore only protons belonging to groups directly bound to the heme iron may experience substantial hyperfine shifts. However, the ZFS produces a detectable pseudocontact shift also for nuclei belonging to non-ligand residues.[18]

The ^{1}H-NMR spectrum of horse aquomet-myoglobin (MetMb-H_2O) dissolved in D_2O and recorded at 400 MHz, and 298 K is shown in Figure 2.

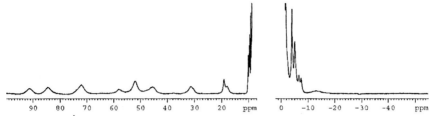

Figure 2 *¹H-NMR spectrum of horse aquomet-myoglobin (MetMb-H₂O) dissolved in D₂O and recorded at 400 MHz, and 298 K*

It represents the spectrum of a typical mammalian ferric high-spin (S=5/2) myoglobin. Four three-proton intensity signals are observed between 92 and 52 ppm. In the region 60-15 ppm many one-proton intensity signals are observed. Few three-proton and one-proton intensity resonances are clearly observable in the upfield part of the spectrum, between –2 and –10 ppm.

While pyrrole substituents show very similar hyperfine shifts for penta- and hexacoordinate high-spin ferric complexes, the meso-H shifts are characteristically downfield (about 40 ppm) for the hexacoordinate systems (horse heart MetMb-H2O) and upfield (-20 to –70 ppm) for the pentacoordinate systems (*Aplysia limacina, Galeorhinus japonicus* and elephant MetMbs).[19]

4.2 Low-spin iron(III) in MetMb-CN

Heme proteins containing low spin Fe(III) comprise the much studied cyanide adduct of MetMb. Due to the low paramagnetism of these octahedral systems (S=1/2) they represent a class of proteins easily studied by NMR. The electron relaxation mechanisms are quite effective and this accounts for short electron relaxation times (10^{-13}s) and consequent relatively long nuclear T_1 values for protons belonging to amino acids located around the paramagnetic iron ion (only 100-200ms) and to heme methyls (80ms), not so shortened as compared to the same diamagnetic form (below 1s). The 1H-NMR spectrum of metMB-CN from horse heart is reported in Figure 3.

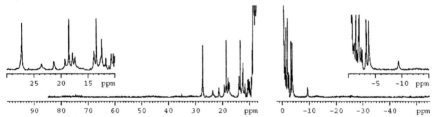

Figure 3 *¹H-NMR spectrum of horse cyanomet-myoglobin (MetMb-CN) dissolved in D₂O and recorded at 400 MHz, and 298 K*

Three methyl signals are observed in the 28-13 ppm region, assigned to 1-, 5- and 8-CH₃ heme groups (see figure 1B).[20] 3-CH₃ is embedded in the diamagnetic region. Of the two paramagnetic contributions to the observed shifts of heme methyls, the contact term dominates. Many one-proton intensity signals are observed in the 24-9 ppm region attributable to the α-type protons of the heme substituents and to the proximal histidine ligand protons. The resonances of some of the protons of other residues present in the active site are also observable due to pseudocontact shifts induced by the magnetic anisotropy of this low-spin system.

The unpaired spin density (ρ, Equation 2) on heme methyl protons depends on the orientation of the axial histidine coordinating the iron ion of myoglobin (Figure 1C). By comparison of the ^{1}H and ^{13}C shift values for the methyl groups of the cyanide adducts of several metmyoglobins from different sources it has been possible to explain their differences in terms of the angle Φ between the projection of the proximal imidazole plane onto the heme plane and the N(II)-Fe-N(IV) vector, and the angle θ_i between the metal-ith-methyl direction and the metal-pyrrole II axis (Figure 1B) according to the equation:[20]

$$\delta_i = 18.4 \sin^2(\theta_i - \Phi) - 0.8 \cos^2(\theta_i + \Phi) + 6.1 \qquad (12)$$

The agreement between the observed and calculated angles is very good and it is possible to use this constrain for structural investigation of low-spin ferriheme proteins.

4.3 High-spin iron(II) in DeoxyMb

Iron(II) in high spin state (S=2) is present in pentacoordinate porphyrin systems. Diamagnetic species are obtained upon dioxygen or carbon monoxide binding in the sixth coordination site. The spectra of DeoxyMB have fewer signals detectable in the portions outside the diamagnetic region. A relatively narrow exchangeable proton was observed at 78 ppm in the ^{1}H-NMR spectrum of sperm whale DeoxyMb and attributed to the solvent exchangeable Hδ1 of the proximal ligand.[21] This signal has been exploited in *in vivo* experiment to assess the formation of deoxymyoglobin in human muscle experiencing hypoxia upon induced ischemia[22] The content of deoxymyoglobin is directly determinable with experiments lasting 15 minutes in spite of its low concentration. Since the Hδ1 signal of the proximal ligand is observable in the ^{1}H-NMR spectrum downfield away from the diamagnetic range it is possible to record selective spectra by positioning the acquisition window far from the water and lipid resonances and by adopting a recycling delay of about 100 ms without compromising the intensity of the signal, which has a very short relaxation time (10ms).

5 PARAMAGNETIC NMR OF FOODSTUFFS

5.1 ^{1}H-NMR of meat extracts from turkey thigh

The ^{1}H-NMR spectra of an aqueous meat extract from turkey thigh meat is shown in Figure 4.

The spectrum of the raw extract, prepared after elution through a cationic exhange column, is characterized by very broad resonances which disappear with both large excess of dithionite (Figure 1C) and slight excess of cyanide (Figure 1D). In addition it is possible to observe some less broad signals falling at chemical shifts values (92, 84, 74 and 53 ppm) corresponding to those assigned to heme methyl in the spectrum of high spin iron(III) of horse heart MetMb-H$_2$O (Figure 2).

It is worth noting that some sharp features (with respect to the other resonances) of the ^{1}H-NMR spectrum are found at around −25 and 32/35 ppm. They disappear for addition of slight excess of either dithionite or cyanide. The signal at -25 ppm could be indicative of a minor species present in solution characterized by pentacoordinated iron(III) more readily reducible and accessible to cyanide ligation. The system is presently under investigation.

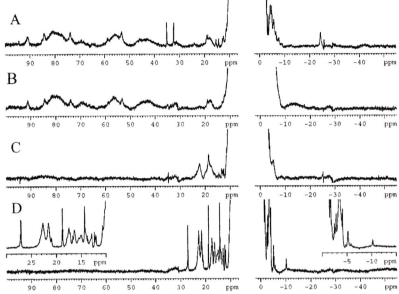

Figure 4 *¹H-NMR spectra of an aqueous meat extract from turkey thigh. A) eluate from the cationic exchange column; B) sample A with addition of 5 mM sodium dithionite; C) sample A with addition of few crystals of solid sodium dithionite; D) sample A with addition of 5 mM sodium cyanide. N.B. The artifacts present in all spectra at 30 and -30 ppm are carrier images.*

5.2 Non heme iron(III): the ovotransferrin seen by the water protons

Hen egg albumen undergoes the thinning phenomenon during the short-term storage (within one week) which bring its viscosity to decrease.[23] The process is traditionally measured by adopting the Haugh index, which is a function of the height of the albumen layer when put on a flat surface.

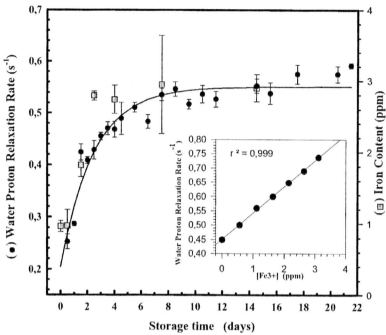

Figure 5 *Water proton longitudinal relaxation rate and iron content measured on hen egg albumen during storage. The inset shows the water proton relaxation rate observed for a solution of ovotransferrin in presence of increasing concentration of iron(III).*

It has also been observed that the water proton longitudinal relaxation rate increases during a comparable time and some studies are proceeding to explain the phenomenon as a consequence of changes in the protein matrix due to the fast pH increase (from 7 to 9 in 12 hours) caused by CO_2 which diffuses away through the shell pores.[24] At the same time, iron(III) flows from yolk to albumen, readily captured by ovotransferrin which prevents its precipitation. The process has the same time scale of the water proton relaxation rate increase, thus suggesting an influence of this paramagnetic ion onto the variation of the NMR parameter (explainable in terms of Equation 11). Figure 5 shows the effect of iron(III) bound to ovotransferrin (present in solution at the same concentration as that found in egg albumen). The iron concentration was varied as it does in egg albumen during the storage: its amount justifies almost 65% of the observed relaxation rate variation in egg, assuming the ion completely bound to ovotransferrin. Further studies aimed at exploring this hypothesis are in progress, based on NMRD profiles.

5 CONCLUSION

High-resolution NMR and low-resolution NMRD techniques, that are consolidated tools to investigate metalloproteins in solution, are shown to hold great potential also in the characterization of foodstuff. The presence of paramagnetic metal ions such as iron(II) and iron(III) induces isotropic shifts and nuclear relaxation that can be exploited to obtain information, e.g., about food ageing. Paramagnetic relaxation effects are also measured on the water pool of foodstuff, and this should be taken as a caveat for the current interpretation of water proton relaxation changes only in terms of changes in hydration and cellular structure of tissues.[25,26]

References

1   I. Bertini, P. Turano and A.J. Vila, *Chem. Rev.* 1993, **93**, 2833.
2   I. Bertini, C. Luchinat and G. Parigi, in *Currents Methods in Inorganic Chemistry* Elsevier 2001, **2**, 1.
3   C.F.G.C. Geraldes and C. Luchinat, *Met. Ions Biol. Syst.* 2003, **40**, 513.
4   L. Banci, I. Bertini and C. Luchinat, in *Nuclear and electron relaxation. The magnetic nucleus-unpaired electron coupling in solution*, Weinheim, VCH, 1991.
5   I. Bertini and C Luchinat, in *NMR of paramagnetic substances*, Coord. Chem. Rev. 150, Amsterdam, Elsevier, 1996.
6   J. Emsley, in *The Elements*, 3rd Ed., Clarendon Press, Oxford, 1998.
7   J.A.T. Pennington, S.A. Schoen, G.D. Salmon, B. Young, R.D. Johnson and R.W. Marts, *J. Food Composition Anal.*, 1995, **8**, 91.
8   C.P. Baron and H.J. Andersen, *J Agric. Food Chem.* 2002, **50**, 3887.
9   S.J. Millar, B.W. Moss and M.H. Stevenson, *Meat Science*, 1996, **42**, 277.
10   D.J. Livingston and W.P. Brown, *Food Technol.*, 1981, **35**, 244.
11   E. Fermi, *Z. Phys.*, 1930, **60**, 320.
12   J.P. Jesson, in *NMR of paramagnetic molecules*, G.N. La Mar, W.D. Jr. Horrocks and R.H. Holm Eds., Academic Press, New York, 1973, 1.
13   N. Bloemebergen, *J. Chem. Phys.*, 1957, **27**, 575.
14   S.H. Koenig, *J. Magn. Reson.*, 1982, **47**, 441.
15   I. Solomon, *Phys. Rev.*, 1955, **99**, 559.
16   S.H. Koenig, *J. Magn. Reson.*, 1978, **31**, 1.
17   G.N. La Mar and F.A. Walker, in *The Porphyrins*, D. Dolphin Ed., Academic Press, New York, 1979, **IV**, 61.
18   G.N. La Mar, G.R. Eaton, R.H. Holm, F.A. Walker, *J. Am. Chem. Soc.* 1973, **95**, 63.
19   U. Pande, G.N. La Mar, J.T.J. Lecomte, F. Ascoli, M. Brunori, K.M. Smith, R.K. Pandey, D.W. Parish and V Thanabal, *Biochemistry* 1986, **25**, 5638.
20   I. Bertini, C. Luchinat, G. Parigi and F.A. Walker, *J. Biol. Inorg. Chem.*, 1999, **4**, 515.
21   G.N. La Mar, D.L. Budd and H.M. Goff, *Biochem. Biophys. Res. Commun*,. 1977, **77**, 104.
22   R. Kreis, K. Bruegger, C. Skjelsvik, S. Zwicky, M. Ith, B. Jung, I Baumgartnerand C. Boesch, *Magn. Reson. Med.* 2001, **46**, 240.
23   D.S. Robinson and J.B. Monsey, *J. Sci. Food Agric.*, 1972, **23**, 893.
24   F.Capozzi, M.A. Cremonini, C. Luchinat, G.Placucci and C.Vignali, *J. Magn. Reson.*, 1999, 138, 277.
25   R.J.S. Brown, F. Capozzi, C. Cavani, M.A. Cremonini, M. Petracci, G. Placucci, *J. Magn. Reson.*, 2000, 147, 89.
26   C. Cavani, M. Bianchi, F. Capozzi, M.A. Cremonini, L. Laghi, M. Petracci, G. Placucci, *Arch. Geflugelkd*, 2002, 66, 165.

INJECTION FLOW NMR AS A TOOL FOR THE HIGH THROUGHPUT SCREENING
OF OILS

S. Rezzi[1]*, M. Spraul[2], D.E. Axelson[3], K. Heberger[1], C. Mariani[4], F. Reniero[1] and
C. Guillou[1]

[1] European Commission, Joint Research Centre, Institute for Health and Consumer
Protection, Physical and Chemical Exposure Unit TP 740, 21020 Ispra (VA), Italy
[2] Bruker Biospin GmbH, Silberstreifen, D-76287 Rheinstetten, Germany
[3] Queen's University, Department of Chemistry, Kingston, Ontario, Canada K7L3N6
[4] Stazione Sperimentale per le Industrie degli Oli e dei Grassi(SSOG), 20133 Milano, Italy

1 INTRODUCTION

Nowadays, there is a strong need for reliable and rapid methods for the control of food oils. The amount of information available in a NMR spectrum and the easy sample preparation render this spectroscopic technique very attractive for the assessment of oil composition. Among the numerous analytical methods developed during the last decade, ^{1}H-NMR spectroscopy of oils demonstrated indeed to provide information, in a rather rapid manner, on lipidic classes, unsaturation level, molar frations of specific fatty acids (linoleic, linolenic acids) and several minor compounds (sterols, squalene, terpenes, oxidixed products...).[1-4] Besides, ^{13}C-NMR gives unique information on the position of fatty acids on glycerol and the stereochemistry of unsaturation.[3-4] More recently, ^{31}P-NMR applied to oils previously derivatized with a phosphorous reagent showed to enable the quantitative analysis of their phenolic compounds and diglycerides.[5-6]

Despite this relatively abundant scientific literature on the NMR applications for oil analysis, until now it has however not been present in control laboratories or used as official method. One of the main reasons for this fact remains the cost of NMR analysis and the know-how required in data acquisition and interpretation. However, the recent developments of flow injection devices allow NMR to emerge as a high throughput system with a high level of automation thus reducing drastically the single analysis cost.[7-8] Besides, the application of pattern recognition statistical methods allow one to process hundreds of spectral information in a multidimensional space offering thus a unique way to vizualize, classify and predict large number of spectra.

Here, we present the application of flow ^{1}H-NMR combined with pattern recognition techniques for the rapid screening of oils from different botanical origins. We also report preliminary results on the differentiation of olive oils according to their year of production.

2 METHOD AND RESULTS

2.1 NMR measurements and statistical analysis

Oil samples were provided by SSOG (Milano, Italy): olive (n=109), peanut (n=4), hazelnut (n=22), colza (n=6), palm (n=6), grape seed (n=4), sesame (n=3), soya (5), sunflower (n=6), corn (n=6).

NMR spectra were measured in the stop-flow mode using a Bruker (Rheinstetten, Germany) DRX-500 instrument operating at 500.13 MHz for ^{1}H observation using a 3 mm single cell ^{1}H/^{13}C inverse detection flow probe with an active volume of 60 μl. Sample transfer from the 96-well plate to the NMR flow probe used a Gilson (Middleton, WI, USA) XL233 automatic sample handling system interfaced to the NMR data system for control and timing (Figure 1).

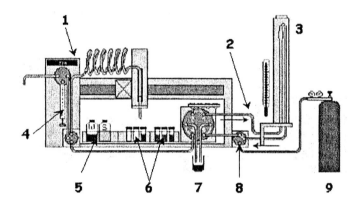

Figure 1 *Direct injection flow-NMR configuration. 1: robot (preparation/injection); 2: transfer line; 3: flow probe; 4: dilutor; 5: solvent rack; 6: samples rack; 7: waste; 8: injection valve; 9: inert gas cylinder for drying (nitrogen)*

For each sample, oil was diluted in 1 mL of deuterated chloroform. ^{1}H-NMR spectra were registered at 300K using 64K data points and the standard Bruker pulse sequence (zgps). The residual CHCl$_3$ signal was suppressed with an off-resonance presaturation generated by a shape pulse of 80ms duration at 7.28ppm. Eight scans were accumulated for each spectrum. The profiles of oils were acquired in 1.30 minute under experimental conditions allowing quantitative measurements. The total experimental time per sample including injection, transfer into the NMR flow-cell, optimisation of the field homogeneity, data acquisition and the inter-sample washing procedure takes around 5 minutes. Under such conditions, one could perform the ^{1}H-NMR analysis of around 300 oils per day. The FIDs were zero filled and Fourier transformed applying a line broadening of 0,1Hz. All spectra were phased manually phase and baseline correction was applied.

Statistical data processing was performed using two methods. The first one consisted in bucketing the ^{1}H-NMR profiles with the AMIX software (Bruker). NMR spectra were reduced in 155 buckets of 0.04ppm width scaled to the total intensity. The bucket areas were then used as variables in the Principal Component Analysis routine of the AMIX software. The second processing method used the whole digitalized data points as input variables for multivariate statistics involving successively Principal Component Analysis and Linear Discriminant Analysis (Statistica).

2.2 Characterization of oils from different botanical origins

Due to its high sensitivity, ^{1}H-NMR is certainly one of the best candidates for a quick assessment of the oil profile. Moreover, ^{1}H-NMR allows one to easily perform both qualitative and quantitative determinations in a relatively reduced experimental time. The time-consuming manual preparation of samples in precision glass NMR tubes, that is required for conventional high-resolution NMR spectroscopy, is clearly a bottleneck for rapid analysis, even if this can be partly improved with the use of sample changers. Nowadays, the recent introduction of flow-NMR represents an interesting alternative that makes it possible to carry out hundreds of ^{1}H-NMR spectra per day with a high level of automation and reproducibility. Flow-probe achieves an automatic direct transfer of a sample into the NMR detector cell itself and adds high-throughput possibilities to the NMR spectrometer.

We were firstly interested in evaluating the performance of the high throughput ^{1}H-NMR profiling combined with pattern recognition techniques for the differentiation of oils from different botanical origins (olive, hazelnut, colza, sunflower, palm, sesame, grape seed, corn, peanut and soya). Even if the compositional differences between such oils can be very important in some cases, a rapid method like flow-NMR combined with pattern recognition techniques could be a reliable tool for eventual adulteration detection and quantification. In order to achieve this, each of the 170 ^{1}H-NMR profiles was subdivided into 155 buckets of 0.04ppm width, which were integrated. The integral values were then processed with Principal Component Analysis (PCA). 154 Principal Components (PCs) were extracted. The first two PCs accounted for 61.7 and 22.9 of the total explained variance, respectively.

A score plot of samples on PC1 versus PC2 is reported (Figure 2). This plot shows that the majority of samples are clearly clustered on PC1 and PC2 according to their botanical origin. The present grouping of samples allowed us to carry out a directional analysis with the loading plot. In that way, a first direction of clustering can be seen from the center to the bottom right corner of the score plot. Indeed, grape seed, sunflower, corn, soya and sesame oils are differentiated from others oils along this direction. The examination of the loadings explaining this clustering allowed us to identify buckets at 2.08ppm and 2.80ppm as most statistically significant signals. These signals represented on Figure 3 arise from the α-olefinic of all unsaturated fatty acids and diacyl protons, respectively. A second direction of clustering can be observed from the center to bottom left corner of the score plot (Figure 2). From the loading factors, we identified the bucket at 1.28ppm as the most significant one, explaining the clustering of olive, peanut and palm oils. This bucket at 1.28ppm covers the methylene protons of all acyl chains. Finally, we can see another direction explaining the clustering of hazelnut and colza oils. The loadings examination indicates the buckets at 2.32 and 5.36ppm as most significant covering the α-carboxyl protons from all acyl chains and the olefinic protons of all unsaturated fatty acids. The spectral regions responsible of the observed clustering on PC1 and PC2 are also reported (Figure 3).

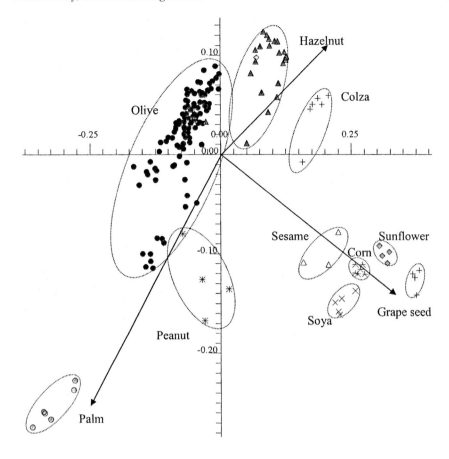

Figure 2 *Score plot of samples on PC1 (horizontal, 61.7% of the total variance) versus PC2 (vertical, 22.9% of the total variance). Main directions from the directional analysis with loadings are represented as arrows.*

These results demonstrate the potentiality of high throughput ^{1}H-NMR combined with pattern recognition technique for the rapid differentiation of different oils. Giving the observed clustering of samples, it seems possible to use this approach to identify, for instance, eventual adulteration of olive oil with other botanical oils.

2.3 Characterization of olive oils according to their year of production

In addition, we were interested in evaluating the relevance of the high throughput ^{1}H-NMR profiling combined with pattern recognition techniques for the differentiation of olive oils from the same geographical origin but produced in different years (2002 and 2003). The possibility to distinguish olive oils according to their year of production using this new method may represent a quick tool for controlling the genuineness of extra virgin olive oils.

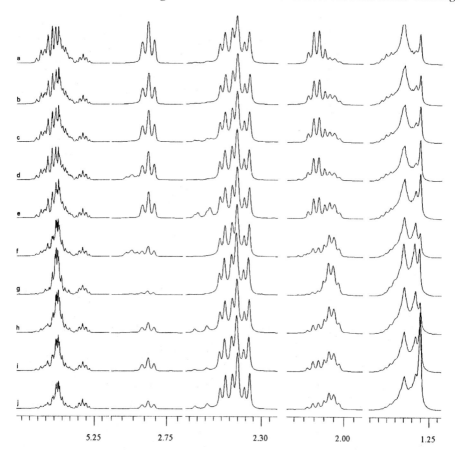

Figure 3 *Spectral regions defined by loadings on PC1 and PC2. Fom left to right: olefinic, diallylic, α-carboxilic, α-olefinic (intensity/2) and methylene (intensity/4) protons. Grapeseed (a), sunflower (b), corn (c), soya (d), sesame (e), colza (f), olive (g), hazelnut (h), peanut (i), palm (j). All signals intensities were calibrated against low-field signal of glycerol at 4.32ppm. Chemical shifts are calibrated against CHCL₃ residue at 7.28ppm.*

Multivariate statistics were used to process the acquired ¹H-NMR profiles. The digitised ¹H-NMR 64K data points were employed as input variables for statistics. Spectral data were firstly reduced by Principal Component Analysis allowing a compression of the total variance of dataset on a reduced number of Principal Components (PCs). The scores obtained on the first 50 PCs explaining all together almost 100% of the total variance were retained for modelling. Linear Discriminant Analysis (LDA) was used as supervised method to classify the samples according to their year of production. LDA selected two and three among the scores as having the largest discrimination power at the 5%

significance level (both for entering and removing variables) for Spain and Greece samples, respectively. Their linear combination (the root or canonical variable) can be seen on the Figures 4 (samples from Greece) and 5 (samples from Spain). It appears that most of the samples can be correctly classified according to the year of production of olive oils from both Spain and Greece. Only two samples, one from Greece (year 2002) and one from Spain (year 2002) were misclassified. These preliminary results showed that the ^{1}H-NMR profile of olive oil can be indicative of the year of production and thus might be relevant for rapid checking of the genuineness of extra virgin olive oils.

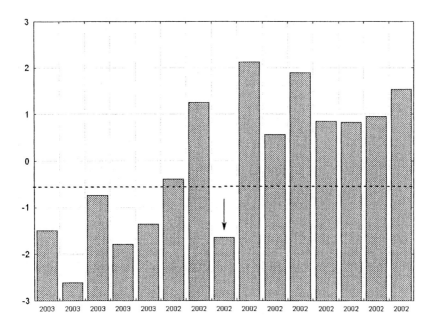

Figure 4   *Classification of olive oils from Greece (n=15) according to the year of production (2002 or 2003). Misclassified case is indicated with an arrow.*

3 CONCLUSION

NMR spectroscopy can provide very important molecular information on the chemical composition of oils. The introduction of flow-NMR represents a real alternative to the conventional high resolution NMR. Flow systems allow NMR to emerge as a high throughput technique as hundreds of spectra can be performed per day. As a result flow-NMR may become a very cost-effective method for food control. Until now the flow-NMR technique has been mainly applied for analysis of biological matrices (biofluids), and here we have also demonstrated its applicability for oil analysis. We processed the ^{1}H-NMR profiles of oils acquired under high throughput mode with multivariate statistics and successfully applied it to the differentiation of oils according to their botanical origin. The application of PCA was sufficient for distinction on the first two PCs of all the botanical origins that were investigated. Besides, the digitized data points of the ^{1}H-NMR profiles were processed successively by PCA and LDA allowing the differentiation of olive oils according to their year of production for samples from Greece and Spain. It appeared thus that Flow-NMR can be successfully applied for the rapid characterization of oils. From another hand, one should also consider the strong potentialities of a combined approach involving a flow-NMR screening of oils with the measurements of the stable isotope by Isotopic Ratio Mass Spectrometry (IRMS) in order to address authenticity of food products in support to European policy for consumer protection.

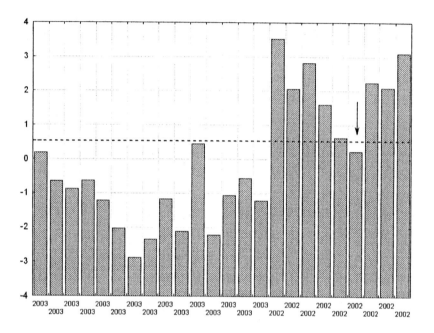

Figure 5   *Classification of olive oils from Spain (n=24) according to the year of production (2002 or 2003). Misclassified case is indicated with an arrow.*

References

1   M.D. Guillén and A. Ruiz. *Trends Food Sci. Technol.*, 2001, **12**, 328.
2   C. Fauhl, F. Reniero and C. Guillou. *Magn. Reson. Chem.*, 2000, **38**, 436.
3   G. Vlahov. *Prog. Nucl. Magn. Reson. Spectrosc.*, 1999, **35**, 341.
4   L. Mannina, A.P. Sobolev and A. Segre. *Spectroscopy Europe*, 2003, **15**, 6.
5   A. Spyros and P. Dais. *J. Agric. Food Chem.*, 2000, **48**, 802.
6   G. Vigli, A. Philippidis, A. Spyros and P. Dais. *J. Agric. Food Chem.*, 2003, **51**, 5715.
7   P. A. Keifer. *Curr. Opin. Chem. Biol.*, 2003, **7**, 388.
8   M. Spraul, M. Hofmann, M. Ackerman, A.W. Nicholls, S.J.P. Damment, J.N. Haselden, J.P. Shockcor, J.K. Nicholson and J.C. Lindon. *Anal. Commun.*, 1997, **34**, 339.

WARPING: INVESTIGATION OF NMR PRE-PROCESSING AND CORRECTION

Frans van den Berg, Giorgio Tomasi and Nanna Viereck

Spectroscopy and Chemometrics Group, Quality and Technology, Department of Food Science, The Royal Veterinary and Agricultural University (KVL), Denmark.
E-mail: fb@kvl.dk

1 INTRODUCTION

NMR is a very all-around analytical technique, suitable for amorphous, heterogeneous and opaque systems such as food and foodstuffs. ^{1}H-NMR is capable of probing both physical and chemical characteristics, allowing food engineers and scientists to obtain a fingerprint of the protons in a sample. Often the ^{1}H-NMR spectrum of a food-system is too complex for direct, deterministic interpretation. Chemometric multivariate factor models like Principal Component Analysis (PCA) can be used to explore highly collinear data set formed by spectral variables, assuming the NMR signals are pre-processed correctly, removing undesired artefacts as source of variation.

NMR signals in the form of frequency domain chemical shift data primarily show the chemical information. Frequency domain chemical shift data are characterized by the fact that molecules, or molecular conformations, give individual chemical shift fingerprints and that the intensity at a given band is proportional to the concentration of the substance, giving rise to that band. The general assumption is that the concentration is proportional to signal intensity and the signals from a given substance remain stable on the frequency axis. Fulfillment of these two assumptions is exactly the requirement of bilinear multivariate models like PCA. However, a number of exceptions to this rule exist: *system-intrinsic* shift by molecular size (e.g. differences in peak position for smaller polymeric molecules and larger polymeric molecules due to their longer T_1 relaxation times), *persistent* or induced chemical shift changes due to pH of the system (e.g. in body fluids), or *noise* shifts (e.g. temperature differences due to sample introduction, magnetic field in-homogeneities or other instrumental or environment effects).

Shift artifacts – possibly important and relevant from a chemical viewpoint – destroy the bilinear nature of NMR data making the factor models unnecessarily complex or even non-interpretable. In this chapter, we will present the blind pre-processing method Correlation Optimized Warping (COW) to handle and correct for them. The problem of shifts and factor modeling will be illustrated via a small data set of four apple juices spiked with different amounts of ethanol.

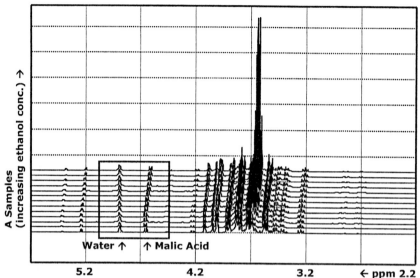

Figure 1 *^{1}H-NMR (raw) spectra of the ethanol standard addition series for apple juice A. Spectra have been baseline lifted to improve visual appearance.*

2 DATA

The example data set used in this chapter consists of ^{1}H-NMR spectra measured on four commercial apple juices (labeled A, B, C and D) [1]. Apple juices A and B were produced by the same manufacturer. Juices A, B and D were produced as an apple concentrate with water added, while C contained pure apple juice. Juices A and C contained unfiltered juice, and B contained pure juice from the apple sort Jonagold. The pH values of the four original apple juices were measured to be A=3.67, B=3.62, C=3.39 and D=3.90. The full ^{1}H-NMR data set consists of 52 spectra of the apple juices each with standard addition of ethanol in 13 levels, between 0.00 and 10.00 % (w/w). The samples were measured at 293K on a 400MHz Bruker Avance NMR spectrometer equipped with an automatic sample changer. Pre-saturation was applied during the 4 seconds relaxation delay to suppress the signal from water protons. The spectra were collected into 32k complex data-points and 128 scans were acquired for each sample. A spectral window of 8278Hz was accumulated in an acquisition time of 1.98 seconds.

The full ^{1}H-NMR spectra of the A apple juice sample series is shown in Figure 1. A clear and systematic shift can be observed as a function of the amount of ethanol added. It can also be observed that the NMR instrument software automatically aligns the water peak indicated in the figure. To get a better view on the problem we will zoom on two identified components (the Water singlet and the Malic Acid duplet) marked in Figure 1. Figure 2a clearly shows that the automatic (approximate) alignment of the water peak and the serious misalignment of the Malic Acid peaks.

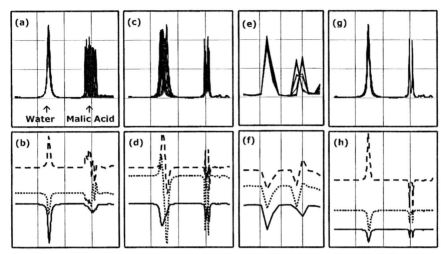

Figure 2 *Zoom region in ^{1}H-NMR spectrum (a) Raw, unprocessed data for all samples; (b) PCA variable loading vectors for Factor 1 (___), 2 (...) and 3 (- -) on raw data for all samples; (c) + (d) shift-corrected data + PCA loadings; (e) + (f) 'bucketing'-corrected data + PCA loadings; (g) + (h) COW-corrected data + PCA loadings*

3 PRINCIPAL COMPONENT ANALYSIS AND THE SHIFT PROBLEM

Principal Components Analysis (PCA) is a popular bilinear factor model used for exploring data tables in chemometrics and many other fields of research [2]. A PCA model finds the best least-squares low rank approximation of a data matrix $\mathbf{X}$

$$\mathbf{X} = \mathbf{t}_1.\mathbf{p}_1^T + \mathbf{t}_2.\mathbf{p}_2^T + \mathbf{E} = \mathbf{T}.\mathbf{P}^T + \mathbf{E}$$
$$\text{minimize}\|\mathbf{X} - \mathbf{T}.\mathbf{P}^T\|^2$$

In these equations object-score vectors fulfill conditions $\mathbf{t}_i^T.\mathbf{t}_i = s_i$ and $\mathbf{t}_i^T.\mathbf{t}_j = 0$ (orthogonal score vectors; size s_i is equal to the square of the singular value), variable-loadings satisfy the criteria $\mathbf{p}_i^T.\mathbf{p}_i = 1$ and $\mathbf{p}_i^T.\mathbf{p}_j = 0$ (orthogonal, normalized loading vectors), and $\mathbf{E}$ is the un-modeled part of $\mathbf{X}$, as illustrated in Figure 3. Hence, the first set of scores and loadings (i.e. the first principal component) is the best approximation of the original data, and the fraction/percentage explained variance captured from the original data-matrix by this first pair expresses how well this approximation succeeded. Similarly, the second score/loading pair is the next best approximation of the original data table. The scores can be seen as new pseudo-values for the objects (^{1}H-NMR samples). The loadings show the role of the original variables (chemical shift axis).

Figure 3 also depicts the shift problem encountered in bilinear models. Looking only at the first vector, a matrix $\mathbf{X}$ is approximated by the outer-product of vectors – scores $\mathbf{t}$ and loadings $\mathbf{p}^T$. The outcome of this decomposition will be the same if entire rows (samples) of columns (variables) are switched (indicated by lines inside matrix $\mathbf{X}$ in Figure 3); the principal components (or eigenvalue structure) of the matrix $\mathbf{X}$ stays intact.

However, if the shift problem occurs the concept of matrix (or a data table in general) is broken: the same column for different samples does not contain the same information. This is visualized by the dots - representing e.g. a peak - in Figure 3. If the dot/peak is shifted in one of the samples it can not be modeled by one single factor. Hence, the score-loading pair *t-p* for the two samples can never show size of this dot/peak in one factor.

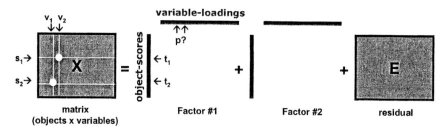

Figure 3 *Conceptual drawing of Principal Component Analysis bilinear factor modelling. Lines in X-matrix indicate sample rows and variable columns; the dots illustrate a shifted peak.*

The practical implication of shift can be seen from our example data in Figure 2b. The water peak – aligned by the NMR software - is well represented by the shape in the loading vectors. For the Malic Acid duplet a severe shift with different degrees of partial overlap between different samples is observed. The first loading vector shows the spread of the *bundle of duplets* from Figure 2a. The second and third factors give a (distorted) picture of the fine structure in this peak bundle. Notice that it is not possible to recognize the duplet in this result. This observation is in agreement with overall conclusion on shift related artifacts in data: factor models become excessively complex because more factors are needed to model phenomena of interest, following a derivative- or Taylor-series like approximation [3].

4 SHIFT-CORRECTION METHODS

In order to compensate for shift-related artifacts we will compare various so-called *blind* pre-processing methods. By blind we mean that no spectral knowledge is used in the correction procedure. Hence, the methods presented do not rely on the (corrected) identification of species in the sample matrix, treating the ^{1}H-NMR measurements as general signal vectors. Many alternative methods for aligning NMR data have been presented in literature; see e.g. [4-6] and references therein. However, many of these methods achieve near-optimal corrections by relying on very specific spectroscopic knowledge in the application phase. In this chapter, we focus on blind methods like COW that achieve non-optimal corrections, but use no domain information and, as such, are often simpler to perform (automatic correction) and are applicable in many different fields of data analysis [3].

By pre-processing we indicate signal correction working on one record/sample at the time. Hence, the operation is optimized based on one signal only (plus a reference or target shared by all samples) and not e.g. on the entire sample set at once, where alignment correction is forms an integrated part of the factor modelling [7].

4.1. Shift correction

The simplest and most straightforward way of aligning two signals is to horizontally shift (left and right direction in a linear fashion) one compared to another and determine the optimal amount of correctional shifting. Two things are important to align a larger sample set: one signal has to be selected as a common, generic reference so that all samples in the data table are aligned towards the same goal; and a criterion for optimality has to be selected. Throughout this work we chose the 5% (w/w) ethanol sample from product A as reference. Although selecting the right reference sample might have a big impact on the results for other pre-processing tasks, the more or less arbitrary choice turned out to be unimportant for all the results presented here. As optimization criterion we selected maximum correlation between samples and reference over a shift range of ±0.048ppm (38 data points). We selected correlation over e.g. a norm-criterion to make the optimization independent of magnitude which is very sample dependent in this ethanol standard addition data set (see Figure 1; total signal length 2771 data points). The outcome for this linear shift correction is shown in Figure 2c-d. In Figure 1 we observed that the shift problem in this data set has two contributions: the main *persistent* effect which is a clear function of the amount of ethanol added, and a smaller random *noise* contribution. The main bulk of the signal (between 3.2 and 4.2ppm) will dominate this linear shift- optimized correction, as observed in Figure 2c. The *persistent* chemical shift is largely corrected for as observed from by Malic Acid duplet, but the water singlet just follows this "correction" leading to a disturbance of that part of the signal. From the PCA loadings in Figure 2d we learn that water peak now shows the characteristic Taylor series or derivative pattern, and the Malic Acid part of the spectrum is not perfectly aligned as a consequence of the *noise* contribution to the shift. Although not perfect when inspecting the results, from Table 1 we learn that the PCA factor model is more parsimonious because the main source of variation in the signal is better aligned.

Table 1 *Cumulative percentages explained variance for PCA models on raw data and alternative pre-processing methods. r^2 is the correlation coefficient between the first PCA factor sample scores and the amount of ethanol added.*

	Factor 1	r^2	Factor 2	Factor 3
Raw data	58%	0.980	70%	80%
Shift correction	88%	0.988	98%	99%
0.05ppm binning	87%	0.965	97%	100%
COW correction	90%	0.988	99%	99%

4.2. Bucketing

Another shift-compensating methods described in literature is to average or integrate the signal over small equidistant intervals of the spectrum to eliminate shift differences between samples within that interval (so-called *bucketing* procedure [8]). The advantage in this approach is that no reference is required, the two main disadvantages are that the method is only semi-blind (placing the intervals is very critical) and the severe reduction in signal resolution by averaging (destroying the fine structure in the spectrum). This can be seen in the result presented in Figure 2e-f where an interval size of 0.048ppm (38 data-points) was used for correction. In Figure 2e it can be seen that the water singlet stays intact in this pre-processing, and a reduction in *noise* shift is observed for this peak. But the shift for the Malic Acid duplet is not captured with this interval setting. It is not possible to find a single equidistant interval correcting for all the shift artifacts observed in Figure 1. Figure 2f illustrates the drop in resolution resulting from this averaging approach: it is not possible to distinguish between the water singlet and the Malic Acid duplet just from the PCA loadings. The percentage explained variance presented in Table 1 for this averaging pre-processing does show a more parsimonious model, but this is a deceiving conclusion. Like PCA factor extraction, the averaging operation is a variance reduction method. Hence, the complexity of the data set (or the effective rank of the matrix in the PCA equation) will be reduced, but this reduction is not necessarily a valuable or compressing reduction corresponding to the aim of e.g. PCA.

Making smaller intervals does not help for the Malic Acid duplet and increasing the interval by one data-point completely merges the water plus Malic Acid information in PCA analysis (not shown). Letting go of the equal size, equidistant requirement would possibly eliminate this problem, but this would of course heavily rely on spectroscopic knowledge.

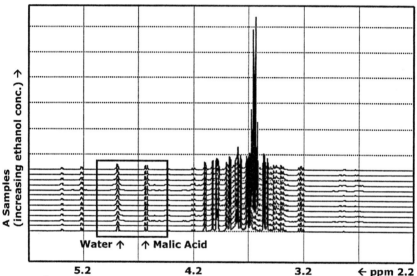

Figure 4 *^{1}H-NMR spectra of the ethanol standard addition series for apple juice A after COW pre-processing*

4.3. Correlation Optimized Warping as Pre-processing

To correct for misalignments in chromatographic data, a procedure called Correlation Optimized Warping (COW) was introduced by Nielsen et al. [9]. It is a piecewise or segmented non-linear pre-processing method aimed at aligning a sample towards a reference by allowing limited changes in all segments lengths on the sample vector. The ratio between the number of points in the reference and the segment length determines the number of connecting segments, or rather the number of segment borders, placed equidistant along the signal. An equal number of connecting segments are specified on the sample vector. The maximum increase or decrease in each sample segment length is controlled by the so-called slack-parameter. The COW-method gets its name from the optimization criterion: the different segment lengths on the sample vector are selected (or the borders are shifted; 'warped') such as to optimize the overall correlation between sample and reference, keeping the segments attached in the original order. The problem can be solved by breaking down the global problem in a segment-wise correlation optimization by means of a dynamic programming algorithm. The solution space of this optimization is limited by two parameters: the number of segment borders, determining the flexibility of the alignment, and the slack range, determining the potential degree of alignment [3, 10]. Figure 2g-h shows the COW pre-processed result for a segment length of 0.0126ppm (10 data-points) with a slack of 0.0025ppm (2 data-points). As can be seen from Figure 2g the non-linear or segment-wise nature of the COW method is able to align the both the water singlet (the *noise* contribution in the shift artifacts) and the Malic Acid duplet (*persistent* contribution) to a high degree. The corresponding loading vectors in Figure 2h clearly show the nature of the signals in the original spectrum. The full COW pre-processed ^{1}H-NMR spectra are shown in Figure 4 (compare to Figure 1).

5 DISCUSSION

Shift artifacts ^{1}H-NMR spectra can limit the application of powerful chemometrics and data analysis methods like PCA factor models. In this chapter we present Correlation Optimized Warping as suitable *blind* pre-processing to handle both *noise* and *persistent* shift contributions. COW can be seen as an intermediate between the established alignment methods (linear) *shift correction* and *bucketing*, incorporating the best of both: correcting locally or segment-wise without loss of resolution.

References

[1] N. Viereck, L. Nørgaard, R. Bro and S.B. Engelsen, Chemometric analysis of NMR data, In Handbook of Modern Magnetic Resonance (Ed. Belton, P.S.), The Royal Society of Chemistry, UK (2004), submitted
[2] I.T. Jolliffe, Principal Component Analysis, 2nd edition, Springer Series in Statistics, Springer-Verlag, New York, USA (2002)
[3] G. Tomasi, F. van den Berg and C. Andersson, Correlation optimized warping and dynamic time warping as preprocessing methods for chromatographic data, Journal of Chemometrics 18(2004)231–241
[4] J.T.W.E. Vogels, A.C. Tas, J. Venekamp, and J. van der Greef, Partial linear fit: a new NMR spectroscopy preprocessing tool for pattern recognition applications, Journal of Chemometrics 10/5-6(1996)425-438
[5] J. Forshed, I. Schuppe-Koistinen and S.P. Jacobsen, Peak alignment of NMR signals by means of a genetic algorithm, Analytica Chimica Acta 487(2003)189-199
[6] R. Stoyanova, A.W. Nicholls, J.K. Nicholson, J.C. Lindon and T.R. Brown, Automatic alignment of individual peaks in large high-resolution spectral data sets, Journal of Magnetic Resonance 170(2004)329-335
[7] B.M. Beckwith-Hall, J.T. Brindle, R.H. Barton, M. Coen, E. Holmes, J.K. Nicholson and H. Antti, Application of orthogonal signal correction to minimise the effects pf physical and biological variation in high resolution ^{1}H NMR spectra of biofluids, The Analyst 127(2002)1283-1288

[8] E. Holmes J.K. Nicholson A.W. Nicholls J.C. Lindon, S.C. Connor, S. Polley and J. Connelly, The identification of novel biomarkers of renal toxicity using automatic data reduction techniques and PCA of proton NMR spectra of urine, Chemometrics and Intelligent Laboratory Systems 44(1998)245-255

[9] N.P.V. Nielsen, J.M. Carstensen and J. Smedsgaard, Aligning of single and multiple wavelength chromatographic profiles for chemometric data analysis using correlation optimised warping, Journal of Chromatography A 805(1998)17-35

[10] Matlab© code for Shift Correction, COW and other alignment methods freely available from www.models.kvl.dk

Applications of Solid-State Methods

MULTIEXPONENTIAL DIFFUSION IN MEAT AT HIGH B-VALUES BY MRI

J.M. Bonny[1], V. Santé-Lhoutellier[2] and J.P. Renou[1]

[1] STIM
[2] BPM - URV INRA 63122 Saint-Genes Champanelle, FRANCE

INTRODUCTION

The extrinsic quantity b characterizes the sensitivity of each NMR sequence to diffusion[1]. Diffusion imaging is generally performed by reconstructing apparent diffusion coefficient (ADC) maps within a low b-value range, assuming a monoexponential relationship between the signal and the b-factor. However, heavily diffusion-weighted experiments using b-values above 1000 s.mm^{-2} have revealed that water diffusion in biological tissues is generally multi-exponential[2]. Hence, the signal intensity can be modeled by

$$S(b;\mathbf{d}) = \sum_i f_i \exp(-b \cdot \mathbf{d}^T \mathbf{D}_i \mathbf{d}) = \sum_i f_i \exp(-b \cdot ADC_i) \qquad (1)$$

where $\mathbf{d}$ is the unit vector defining the direction of measurement, $\mathbf{D}_i$ is the i-th diffusion tensor, and f_i its respective amplitude. Biexponential behavior of diffusion curves was reported in muscle tissue by Henkelman et al.[3] using a non-localized PGSE technique and an extended range of b-factors. The aim of this study was first to assess the probable biexponential behavior of diffusion curves in meat by MR imaging at 4.7 T. Diffusion tensor imaging experiments were also performed using both variable b-factors and six non-collinear diffusion gradient directions. This was designed to assess the diffusion model in terms of two tensors and to quantify new parameters completing the analysis of muscle structure previously performed using a single tensor model at low b-values[4].

2 METHODS AND RESULTS

2.1 Multiexponential analysis of diffusion curves by MRI

2.1.1 NMR measurements. *Biceps femoris* (n = 2, BF), *Infraspinatus* (n = 1, IS), *Longissimus thoracis* (n = 1, LT), and *Pectoralis profundus* (n = 1, PP) muscles of Charolais cows, aged 3 to 7 year old, were chosen for their different architectures, including fiber bundle size, intramuscular connective thickness and hierarchy. Samples were aged 20-24 days *post mortem*. Each sample was positioned so that the main muscle fiber direction was approximately parallel to the main magnetic field direction.

Each sample was placed in a birdcage probe for microimaging at room temperature (~20°C). An axial single slice of 10-mm thickness was analyzed with a 128 x 128 matrix size and a field of view of 64 x 64 mm^2, giving a voxel volume of 0.5 x 0.5 x 10 = 2.5 mm^3. Diffusion-weighted images were recorded using a pulsed-gradient spin-echo sequence ($\delta = 9.5$ ms, $\varDelta = 17.5$ ms). The other parameters were: TR/TE = 3000/38.9 ms, 195 Hz per pixel readout bandwidth, Nex = 4. The diffusion gradient amplitude was increased in order to obtain a linear progression of the 16 *b*-values over an extended range of up to 10000 s/mm^2 in a single diffusion gradient direction ($[1,1,1]^T$ in the image referential). As defined by Meier et al.[5], this corresponds to a constant diffusion time experiment (*ct*-experiment) with $t_D = \varDelta - \delta/3 = 14.3$ ms.

2.1.2 Image analysis. To assess the possible multiexponential behavior of diffusion decay curves in a given tissue type, i.e. myofibers and intramuscular fat (IMF), there are two possible drawbacks to take into account: the effect of asymmetric Rician noise which biases the parameters of curves fitted on magnitude samples of low SNR (< 5), and the partial volume effect which may mix the signals coming from myofibers and IMF. To prevent these effects, the DW curves were fitted using a multiexponential model only in the voxels presenting an SNR above 5, and the diffusion-weighted image obtained at $b = 10000$ s/mm^2 was thresholded in order to extract voxels belonging to the unmixed myofibers class (figure 1).

Biexponential decomposition of DW decay curves was performed using the non-linear least-squares Levenberg-Marquardt algorithm. An F-test was used to assess whether the biexponential model could be considered as further improving the χ^2 misfit compared to the monoexponential model. It was judged to improve fitting only if the probability that the noise accounted for the misfit improvement when comparing χ^2(*mono*) and χ^2(bi) was less than 1%.

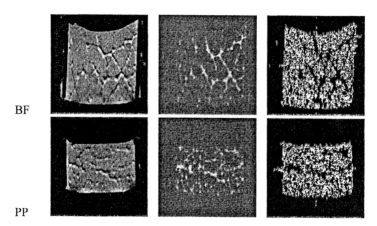

Figure 1   *Diffusion-weighted images of BF and PP (left) obtained with b = 100 s/mm^2 (center) and b = 10000 s/mm^2 (right). Binary image showing voxels of pure myofibers. It is obtained by thresholding the Diffusion-weighted images obtained at b = 10000 s/mm^2 from SNR = 5 to 15.*

2.1.3 Results. Figure 2 illustrates the large differences in DW decay curve characteristics between IMF and myofibers. Both demonstrate biexponential behavior; in myofibers $ADC_f \sim 1.20 \ 10^{-3} \ mm^2/s$ and $ADC_s \sim 0.07 \ 10^{-3} \ mm^2/s$, the fast component being the largest one $f \sim 96\%$. This is highly consistent with the values reported by Henkelman et al.[3] obtained at a comparable temperature on bovine muscle samples analyzed 2 hours *post mortem* ($ADC_f = 1.18 - 1.43 \ 10^{-3} \ mm^2/s$, $ADC_s = 0.07 \ 10^{-3} \ mm^2/s$ and $f = 96\%$).

In the IMF, the two diffusion components probably result in contamination of the myofibers signal due to partial volume effect (favored by the thick slice used) and chemical shift artifact. This observation is supported by ADC_f values in myofibers and IMF which are largely the same. When the myofiber contamination is reduced by the attenuation of the fast bulk component for $b > 4000 \ s/mm^2$, the remaining slow component highlights a slow ADC_s of $\sim 0.01-0.02 \ 10^{-3} \ mm^2/s$. The latter coefficient is consistent with the values reported recently by Lehnert[6] in the subcutaneous fat of human subjects. It underlines the reduced mobility of fat protons and explains the predominence of fat protons on heavily DW images (figure 1), which may be useful for generating high resolution images of IMF as discussed above.

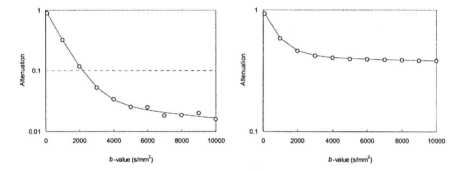

Figure 2 *DW decay curves of (left) myofibers and (right) IMF. The biexponential model is given as a solid line.*

2.2 Multitensor analysis of diffusion by MRI

2.2.1 NMR measurements. The two *Pectoralis* muscles from the same turkey were studied and compared within each experiment. After slaughter, the first muscle (reference sample) was simply maintained at 10°C after excision until MR experiment 24 h *post mortem*. The second muscle (exudative sample) was maintained at high temperature (42°C) in order to reproduce pale, soft and exudative (PSE) meat quality defect in the turkey, which is characterized by denatured proteins and therefore by high water loss from the meat[7]. The two samples were positioned so that the muscle fiber main direction was approximately parallel to the main magnetic field direction. The two samples were placed in the microimaging birdcage at a constant temperature of 10°C. A single slice of 10-mm thickness was analyzed with a 128 x 64 matrix size and a rectangular field of view of 128 x 64 mm^2, giving a voxel volume of 1 x 1 x 10 = 10 mm^3. Diffusion-weighted images were recorded using a pulsed-gradient spin-echo sequence ($\delta = 25$ ms, $\Delta = 30$ ms, $t_D = 21.7$ ms). The other parameters were: TR/TE = 700/62.5 ms, 156 Hz per pixel readout bandwidth, Nex = 2.

A *ct*-experiment with $t_D = 21.7$ ms was performed with a linear progression of the 16 *b*-values over an extended range of up to 5000 s/mm². For each *b* value, six non-collinear diffusion gradient directions were applied, which is sufficient for estimating the diffusion tensor[8]. Hence, the total acquisition protocol generates 96 diffusion-weighted images.

2.2.2   Image analysis.   Although biased in comparison to full fitting approaches, one robust approach consists in estimating monoexponential tensors $\mathbf{D_f}$ and $\mathbf{D_s}$ from the early rapidly ($b \sim$ 300-1250 s/mm²) and later slowly ($b \sim$ 4000-5000 s/mm²) decaying magnitude. This approximation was performed by solving an overdetermined linear system using a multivariate linear regression model[9]. Possible contamination by the intramuscular fat (IMF) was not accounted for due to the very low amount IMF in turkey meat.

2.2.3   Results. $\mathbf{D_f}$ and $\mathbf{D_s}$ were first of all characterized by significant differences in mean diffusivity ($0.75 \cdot 10^{-3}$ mm²/s and $0.45 \cdot 10^{-3}$ mm²/s, respectively), the fast tensor being the bulk fraction f $\sim$ 69%. In spite of the effect of noise which induces an overestimation of the anisotropy, especially for $\mathbf{D_s}$, both tensors demonstrated a significant anisotropy; respectively 0.20 and 0.25 of fractional anisotropy index. This is in agreement with previous studies performed on other biological tissues[10, 11] which also reported a highly anisotropic slow tensor. Since the diffusion process is anisotropic in the two regimes, it was relevant to compare the directions of principal diffusivity of each tensor which are given by their first eigenvectors. It appears that these directions are largely the same.

When normal and PSE meats are compared, there is only a slight reduction of anisotropy of the fast bulk tensor $\mathbf{D_f}$. This is consistent with the exudation phenomenon in PSE meat which produces a greater fraction of isotropic free water[4].

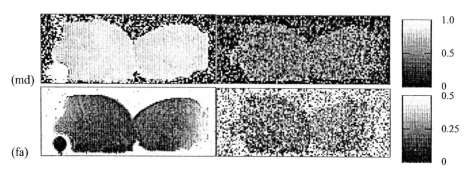

Figure 3   *Parametric maps showing mean diffusivity (md) and fractional anisotropy (fa) for the fast (left) and the slow (right) apparent tensors*

2.3 Imaging of fat using diffusion contrast at high b value

2.3.1 NMR measurements. At room temperature ADC of fat is at least five times lower than the bulk ADC of myofibers. A PGSE reference acquisition (nil *b*-value) of a bovine meat fatty sample was first performed. Using this reference, the *b* factor was determined in order to attenuate the water signal (i.e. signal-to-noise ratio below 1) by accounting for the concomitant effects of voxel volume change and of the attenuation due to diffusion.

A 3D-PGSE sequence was implemented which included a composite non-selective pulse for limiting the refocusing errors due to B_1 inhomogeneities over the whole field of view.

For an *Infraspinatus* bovine muscle at ~ 20°C and according to the measured ADCs, the adequate *b*-value was equal to 1200 s/mm^2 in the case of an isotropic acquisition with a voxel volume of 0.064 mm^3. The latter corresponds to a field of view of 51.2 mm and a matrix size of 128, giving an acquisition time of 28 min (TR = 100 ms).

2.3.2 Results. 3D-PGSE images of the muscle sample are shown in figure 4. Because of the high spatial resolution and the voxel isotropy, intramuscular fat distribution is depicted with the same quality whatever the incidence. For example, the continuity of fat network along the muscle fascicles is clearly demonstrated. Compared to previous techniques proposed for generating images of fat distribution, the PGSE technique requires neither careful adjustment of inversion time (like the water suppressed T1 weighted method) nor high-order correction of B_0 inhomogeneity (like the CHESS and the Dixon phase-sensitive techniques). PGSE is especially well dedicated to a high-resolution depiction of fat distribution because the b-value needed to attenuate the myofiber signal decreases with the voxel volume. A drawback of this method is the sensitivity of DW images to bulk motion, limiting its use to motionless samples. Furthermore, any factor inducing a variation of the ADC contrast between fat and myofibers modifies the parameter set-up (e.g. temperature, drying).

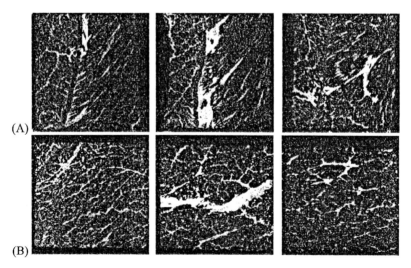

(A)

(B)

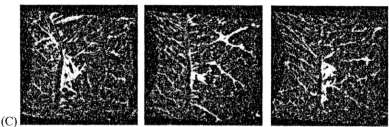

(C)

Figure 4   *Successive planes (from left to right, spacing of 12.8 mm) obtained from the*
3D-PGSE block of the Infraspinatus bovine muscle reconstructed in three
orthogonal incidences (A, B, C)

3 CONCLUSIONS

Both multiexponential and multitensor analyses have demonstrated that the diffusion process in meat is multicompartmental at voxel volume level (i.e. 2.5 – 10 mm^3). The measurement of the diffusion parameters characterizing such processes requires an extended range of b factors up to 5000 s/mm^2 for resolving at least two components. Hence, efficient gradient coils and high signal-to-noise ratio are needed because the available signal is rapidly attenuated by diffusion at a high b value (i.e. ~ 250-300 at b = 10000 s/mm^2).

An important consequence of our results is that, except for the direction of principal diffusion which is an invariant, other characteristics given by the apparent diffusion tensor depend on the b range selected. Therefore, characterization of the tissue structure using multiple tensors first requires being able to fit the complicated multitensor model given by eq. (1) with robustness.

Finally, 3D-PGSE-based imaging of fat-may have useful implications, since fat influences some important meat quality characteristics, e.g. marbling, sensory properties and keeping qualities as well as the diffusion of water and salt in cured/smoked products.

References
1. Le Bihan, D., E. Breton, D. Lallemand, P. Grenier, E. Cabanis, and M. Laval-Jeantet. *Radiology* **1986**, 161, 401.

2. Pfeuffer, J., S. W. Provencher, and R. Gruetter. *MAGMA* **1999**, 8, 98.

3. Henkelman, R. M., G. J. Stanisz, J. K. Kim, and M. J. Bronskill. *Magn. Reson. Med.* **1994**, 32, 592.

4. Bonny, J. M., and J. P. Renou. *Magn. Reson. Imaging* **2002**, 20, 395.

5. Meier, C., W. Dreher, and D. Leibfritz. *Magn. Reson. Med.* **2003**, 250, 500.

6. Lehnert, A., J. Machann, G. Helms, C. D. Claussen, and F. Schick. *Magn. Reson. Imaging* **2004**, 22, 39.

7. G. Monin, "Conversion of Muscle to Meat : Colour and Texture Deviations" in *Encyclopedia of Meat Science,* 2004, Elsevier, 323.

8. Basser, P. J., and C. Pierpaoli. *Magn. Reson. in Med.* **1998**, 39, 928.

9. Clark, C. A., M. Hedehus, and M. E. Moseley. *Magn. Reson. Med.* **2002**, 47, 623.

10. Hsu, E. W., D. L. Buckley, J. D. Bui, S. J. Blackband, and J. R. Forder. *Magn. Reson. Med.* **2001**, 45, 1039.

11. Maier, S. E., S. Vajapeyam, H. Mamata, C. F. Westin, F. A. Jolesz, and R. V. Mulkern. *Magn. Reson. Med.* **2004**, 51, 321.

STUDY OF FAT AND WATER IN ATLANTIC SALMON MUSCLE (*SALMO SALAR*) BY LOW-FIELD NMR AND MRI

E. Veliyulin, I. G. Aursand and U. Erikson

SINTEF Fisheries and Aquaculture, 7465 Trondheim, Norway

1 INTRODUCTION

Development of non-destructive methods for studying the interactions of water and fat with the structure changes occurring during fish processing may provide the insight necessary to improve the quality of such products.

Various modalities of Nuclear Magnetic Resonance (NMR) offer several non-destructive applications that can provide with versatile information about the structure of various biological systems. Low-field (LF) NMR has mainly been used for relaxation time studies and quantification of various components such as water and fat in foods. The microscopic structure of biological systems consists of a network of macromolecules that interact with water protons. Also, water is physically localized in various compartments in the tissues. ^{1}H NMR relaxation spectra of such systems may be complicated as the NMR responses from different ^{1}H pools are usually observed simultaneously and it is not always possible to separate relaxation contributions originating from different pools or different substances (for example fat and water). Interpretation of the NMR relaxation spectra is still a matter of discussion due to the complexity biological systems. For instance, in pork the simple intra-/extracellular compartmentalization theory suggested earlier[1] could not satisfactory explain all features of the multiexponential transverse relaxation. Three commonly observed relaxation components are attributed to the water tightly associated with macromolecules (the fastest relaxation component at 1–10 ms), water located within highly organized protein structures (the intermediate component at 40-60 ms), and water between fiber bundles (the slowest component at 150-400 ms)[2]. In fatty fish such as Atlantic salmon, the interpretation of relaxation spectra is complicated by the fact that the fat relaxation components interfere with those of water[3].

Magnetic Resonance Imaging (MRI) is a technique that offers a unique opportunity to produce cross-section images of intact whole fish. Depending on the particular task, MRI instruments can produce different types of visible contrast in the MR images. This is achieved by programming and running specific MR sequences that can differentiate the NMR response of the protons localized in molecules with different mobility or chemical environment. For example, it is possible to obtain MR images of 'water' and 'fat'[4], 'diffusion weighted' images where only molecules with low mobility are visible[5] or high resolution images of connective tissue[6]. A newly developed method based on double-quantum filtered MRI detects only molecules associated with ordered tissue structures, suppressing the signal from isotropic fluids[7]. MRI is also a powerful technique to visualize and monitor various dynamic processes, allowing to dynamically follow processes non-destructively and with high spatial resolution. For instance combined with NMR spectroscopy MRI can be a valuable tool for studies of fresh and frozen fish[8].

The goal of the present study was to separate fat and water in fatty fish by: (1) showing the advantages of the 2D diffusion weighted T_2 relaxation method compared with the conventional 1D relaxation method, and (2) to develop MRI protocols to produce separate, quantitative 'fat' and 'water' images.

2 EXPERIMENTAL

2.1 LF NMR: Diffusion weighted transversal relaxation time studies

A farmed Atlantic salmon (*Salmo salar*) fillet was bought at the local fish market three days post mortem. Three parallel samples of the white muscle close to the belly flap area were stamped out of the fillet using a specially designed coring tool. Then the samples were transferred to NMR tubes (10 mm in diameter). The tube filling height was about 1 cm and the approximate sample weight was 0.4 g. After the LF NMR measurements, the NMR tubes containing samples were frozen at –25 °C (24 h) and thawed at 5 °C before repeating the measurements at 10 and 25 °C.

The LF measurements were performed using the minispec mq NMR analyzer (Bruker Optik GmbH, Germany) with a magnetic field strength of 0.47 Tesla corresponding to a proton resonance frequency of 20 MHz. The instrument is equipped with gradient coils producing magnetic field gradients of up to 3.2 Tesla/m. A water bath (Haake UWK 45, Germany) was connected to the probehead to make measurements at 10 or 25 °C regulating the sample temperature with an accuracy of 0.1°C. Before measurements, all samples were thermostated to 10 or 25 °C for 1 hour in a separate water bath (Julabo F10, Germany).

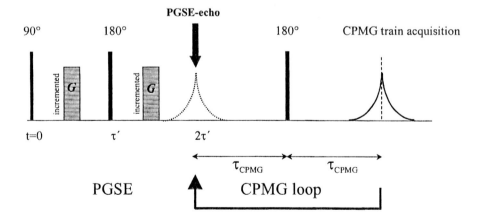

Figure 1   *The PGSE-CPMG pulse sequence for 2D diffusion vs. transversal relaxation time studies*

To obtain a two-dimensional data set of diffusion weighted transversal relaxation curves, a modification of the well-known Carr-Purcell-Meiboom-Gill (CPMG) pulse sequence was used. A pulsed field gradient spin echo sequence (PGSE) was combined with a train of 180° refocusing pulses (Figure 1). The amplitude of the PGSE gradients was incremented in steps (0.16 Tesla/m) from 0 to about 3.2 Tesla/m, increasingly suppressing the contribution of the most mobile components to the resulting echo. Corresponding relaxation curves were acquired at each gradient step. The following acquisition parameters were used: echo-time (TE) of the PGSE part of the sequence was set to 20 ms, duration of the gradient pulses was 1 ms, TE of the CPMG train was 0.2 ms, relaxation delay (RD) was 2 s and 4000 even echoes were acquired in 8 scans. Thus, a 2D data set with 4000 rows (dimension of relaxation) and 21 columns (dimension of diffusion) was obtained for each sample. In addition, conventional CPMG relaxation curves were measured for all samples with the same acquisition parameters as the CPMG echo train in the 2D experiment.

The 2D data were processed by the newly developed 2-D Inverse Laplace Transform[9], using software package[10] implemented in MatLab (The MathWorks, Inc., USA). The same software was used for processing of the 1D relaxation curves.

2.2   MRI: Quantification of white muscle fat and water

The MRI studies were performed using a Bruker Avance DBX100 instrument (Bruker BioSpin, Germany). The instrument has a horizontal wide bore opening suitable for imaging of comparatively large objects (^{1}H imaging area - sizes up to 15 cm in diameter and 15 cm in length).

The chemical shift selective (CHESS) imaging protocol makes use of the fact that the protons in fat and water molecules have slightly different NMR resonance frequency. Hence they have different chemical shifts. When using the technique, a special frequency selective RF pulse with a predefined bandwidth is applied to excite either the fat or the water component only. This makes it possible to achieve an acceptable degree of separation. Thus, separation of fat and water in MR images can be achieved.

A piece of frozen-thawed Atlantic salmon white muscle (approximately $3 \times 3 \times 4$ cm) was placed in the iso-center of the magnet. In addition, two reference samples containing 100 % fish oil and 100% distilled water were placed within the imaged area. The water reference was doped with 0.0001 m/l $MnCl_2$ to shorten the proton relaxation time to $T_2 = 260$ ms. Three types of MRI images were acquired: 'a proton density' image, 'a fat image' and 'a water image'. The following acquisition parameters were used: RD = 2 s, number of excitations (NEX) = 40, field of view (FOV) = 5 x 5 cm and TE = 16.2 ms (fat images) and 26.4 ms (water images). Images of 20 slices with a slice thickness of 1.5 mm were acquired. For fat excitation, a selective sinc-shaped pulse with a bandwidth (BW) of 700Hz was centered (CF) -600 Hz away from the water frequency to avoid signal contribution from water. For water signal excitation, a similar sinc pulse with a bandwidth of 480 Hz was centered +200Hz away from the water resonance frequency to avoid signal contribution from the fat (Figure 2). The 'proton density' image was acquired using a Multi-Spin Multi-Echo (MSME) protocol with TE = 4.1 ms, TR = 2 s and NEX = 4.

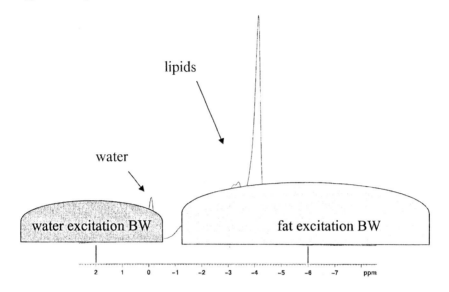

Figure 2 *Selective water (CF = +200 Hz, BW =480 Hz) and fat (CF = -600 Hz, BW = 700 Hz) excitation in the CHESS MRI protocols*

The obtained 'fat' and 'water' images were quantified with an in-house made MRI software package using the Interactive Data Language (Research Systems Inc., UK). Prior to calculation, the original images were scaled in the intensity range from 0 to 255. A histogram of all pixel intensities from the 'water' image is shown in Figure 3a. In this histogram the first peak (approximate intensity range 0 – 43) represents the noise in the image, the second peak (intensity range 43 – 150) originates from the salmon sample and the third one (intensity range 150 – 255) originates from the water reference. In order to quantify the water content, the 'noise' peak was first filtered out using the minimum between the 'noise' and the 'water' peaks (see Figure 3a) as a cut-off value. After that, the average intensity of the reference oil sample was set to 100 and the rest of the image pixel intensities were rescaled accordingly, resulting in a histogram as shown in Figure 3b. When calculating the mean intensity for all sample pixels of the modified image, the water content in the corresponding MRI slice could be determined. By averaging the water content of all slices (covering practically the whole sample), the mean water content of the whole sample was determined. Since the T_2 relaxation time of the sample ($\approx$ 44 ms) was different from that of the water reference (260 ms), the calculated intensities had to be corrected correspondingly. An identical algorithm was used to determine the sample fat content. The programmed automated calculation routines could calculate fat and water contents in all image slices in about 20 s. After the MRI experiment, the salmon sample was divided in two equal parts, one for fat content determination by chemical extraction[11], while the other part was used for total water content determination by weight change after drying at 105 °C for 24 h.

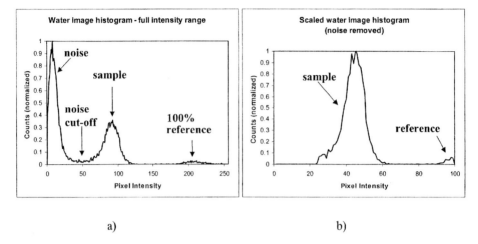

a)   b)

Figure 3   *a) Pixel intensity histogram as acquired from the original 'water' image;*
b) Same histogram after noise filtering and rescaling

3 RESULTS AND DISCUSSION

3.1 Diffusion weighted transversal relaxation time studies

Figure 4 shows diffusion vs. T_2 relaxation time distribution maps for fresh (10°C) and frozen-thawed muscle at 10 or 25 °C, respectively. The corresponding 1D T_2 distributions are the shown above the maps. From the 2D distribution maps the relaxation times and corresponding diffusion constants for the observed peaks could be roughly estimated. A water peak ($T_2 \approx 44$ ms, $D \approx 1.0\times10^{-9}$ m^2/s) and a fat peak ($T_2 \approx 100$ ms, $D \approx 2.7\times10^{-11}$ m^2/s) were observed in the fresh sample at 10°C. No clear changes were observed between fresh and frozen-thawed tissues when measured at 10°C. An advantage of the 2D distribution map approach compared with the conventional 1D is seen by inspecting the longer relaxation peak ($T_2 \approx 140$ ms) from the frozen-thawed sample measured at 25 °C. While the respective water and fat peaks completely overlap on the 1D distribution, the 2D map clearly shows the presence of the two phases at this relaxation time. At relaxation times about 140 ms, the diffusion constants at 25°C were 2.0×10^{-9} (water) and 2.6×10^{-10} m^2/s (fat), respectively. The water component with the longer relaxation time corresponds to that observed in cod[12,13] and pork muscle[2,14], which is thought to represent extra-myofibrillar water. After increasing the temperature from 10 to 25 °C a shift towards slightly shorter relaxation time and higher diffusion constant was observed for the major water component ($T_2 \approx 37$ ms, $D \approx 2.0\times10^{-9}$ m^2/s). In a marine coldwater species such as Atlantic salmon, the tissues contain polyunsaturated fatty acids. This fraction of the total fat content is mobile even at very low temperatures. Therefore, more research is needed to elucidate the "NMR-behavior" of this fat as seen in relation to changes in the extramyofibrillar water pool.

        The technique may be improved in order to further minimize the effect of Eddy-currents on the observed CPMG echo train. This would allow running the PGSE pulse sequence at substantially shorter echo times making it possible to obtain information about the fast relaxing components such as water bound to macromolecules and immobile fat. Further development of the data processing software would allow separate quantification of various water pools and lipids by 2D integration of the corresponding peaks. The method can be valuable for investigation of several types of complex food systems, detecting the presence of different water and fat phases in the sample. Furthermore, 2D diffusion weighted relaxation experiments performed under different food processing conditions (temperature, pressure, humidity, salt content etc.) may be useful to provide a better understanding of the system. In turn, this information can be used to optimize various food unit operations.

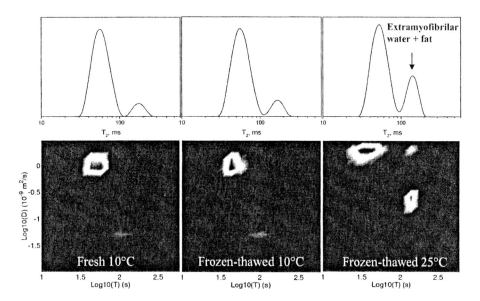

Figure 4   *Diffusion vs. T_2 relaxation time distribution maps (image resolution: 15×15 pixels) for fresh ($10\,°C$) and frozen-thawed (10 and 25 °C, respectively) Atlantic salmon muscle. Corresponding 1D T_2 distributions are shown at the top of each map.*

3.2 Quantification of fat and water by MRI

Figure 5 shows three types of MRI images acquired in the experiment: 'proton density' image (a), 'fat image' (b) and 'water image' (c). In the 'proton density' image both fat and water components exhibited their maximal intensity. The doped water reference was almost invisible on the 'fat image' - and vice versa - the fat reference was almost invisible on the 'water image' proving a high degree of separation of fat and water. Satisfactory suppression of either the fat or water spectral component could only be achieved in a region with a highly homogeneous magnetic field and RF pulses. Therefore, the MRI instrument had to be carefully shimmed prior to image acquisition. An even better separation between fat and water could be expected when using an NMR instrument with a higher magnetic field due to the increased spectral separation in the frequency domain between fat and water.

The water and fat contents were calculated from the respective MRI images. The average of all slices in case of water and fat were 57 ± 3 % and 19 ± 2 %, respectively. The corresponding values obtained from chemical extraction-based analyses of the same sample were 59 % and 21 %, i.e. within the estimated error of the MRI method. Slightly lower MRI values of both fat and water may be explained by partial suppression of the signal from the observed component and by the roughness of the cut-off algorithm for noise filtering. A possible improvement of the noise filtering method should include extrapolation of the left shoulder of the main peak (see the histogram in Fig. 3b) by fitting it with an appropriate model function.

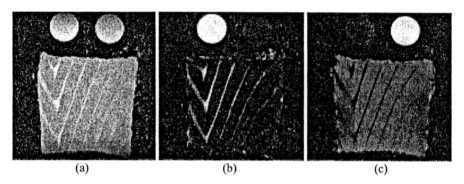

(a)   (b)   (c)

Figure 5 *Three types of MR images: a) 'proton density' image, b) 'fat image', and c) 'water image'. Two NMR tubes filled with $MnCl_2$-doped distilled water (upper right) and 100% fish oil (upper left) were used as references.*

4 CONCLUSIONS

Compared with conventional 1D relaxation studies, a 2D diffusion weighted relaxation experiment can provide with additional information regarding the presence and mobility of different compounds in biological tissues. At 10°C, we did not observe significant differences in the 2 D diffusion-relaxation map when fresh and frozen-thawed Atlantic salmon muscle was compared. Subsequent heating to 25 °C revealed an additional water component with approximately the same relaxation time as the fat component.

A Chemical Shift Selective MRI protocol (CHESS) was successfully applied to produce separate water and fat images of an Atlantic salmon white muscle. The images were quantified using a simple intensity histogram-based approach. However, more advanced MRI protocols should be evaluated for possible reduction of the magnetic field and RF pulse inhomogeneity effects.

Acknowledgements

Financial support of the Research Council of Norway is gratefully acknowledged. Many thanks to Research Engineer Trond Singstad (St.Olavs Hospital, MR Center, Trondheim, Norway) for the technical support and help with optimization of MRI protocols.

References

1 W.C. Cole, A.D. Le Blanc and S.G. Jhingran, *Magn. Reson. Med.*, 1993, **29**, 19
2 H.C. Bertram, A.H. Karlsson, M. Rasmussen, O.D. Pedersen, S. Dønstrup and H.J. Andersen, *J. Agric. Food Chem.*, 2001, **49**, 3092
3 E. Veliyulin, C. van der Zwaag, W. Burk and U. Erikson, *"In-vivo determination of fat content in Atlantic salmon (Salmo salar) with a mobile NMR spectrometer"*, Accepted to *J. Sci. Food Agric.*, 2004
4 W. Laurent, J. M. Bonny and J.P. Renou, Quantification of fat and water fractions in high field MRI using a multislice chemical shift selective inversion recovery (CSS-IR). *Proceedings of XVIII-th International conference on Magnetic Resonance in Biological Systems*, Tokyo, Japan, 1998, IV (9), 88.
5 R.V.Mulkern and R.G.S. Spencer , *Magn. Reson. Imaging* , 1988, **6**, 623
6 J.M. Bonny, W. Laurent, R. Labas, R.G. Taylor, P. Berge and J.P. Renou, *J. Sci. Food Agric.*, 2001, **81**, 337
7 L. Tsoref, H. Shinar, Y. Seo, U. Eliav and G. Navon, *Magn. Reson. Med.*, 1998, **30**, 720
8 N. Howell, Y. Shavila, M. Grootveld and S. Williams, *J. Sci. Food Agric.*, 1996, **72**, 49
9 M. Hürlimann and L. Venkataramanan, *J. Magn. Reson.*, 2002, **157**, 31
10 S. Godefroy, B. Ryland and P.T. Callaghan, *2D Laplace Inversion*, Victoria University of Wellington, New Zealand, 2003
11 E.G. Bligh and W.J. Dyer, *Can. J. Biochem. Physiol.*, 1959, **37**, 911
12 U. Erikson, E. Veliyulin, T. Singstad and M. Aursand, *J. Food. Sci.*, 2004, **69**, 107
13 C. Steen and P. Lambelet, *J. Sci. Food Agric.*, 1997, **75**, 268
14 H.C. Bertram, H.J. Andersen and A.H. Karlsson, *Meat. Sci.*, 2001, **57**, 125

SLOW-MAS NMR METHODS TO STUDY METABOLIC PROCESSES *IN VIVO* AND *IN VITRO*

Robert A. Wind[1], Hanne Christine Bertram[2], and Jian Zhi Hu[1]

[1] Pacific Northwest National Laboratory, Richland, WA, USA
[2] Danish Institute of Agricultural Sciences, Department of Food Science, Research Center Foulum, Tjele, Denmark

1 INTRODUCTION

Metabolism involves the cell- and organism-level chemical processes that are necessary for life. A metabolite is any substance participating in metabolism, either as a precursor, intermediate, catalyst or product. Metabolites include a large range of substances such as organic acids, carbohydrates, amino acids, nucleosides, peptides, and so on. Metabolic processes depend upon the metabolite concentrations (profiles), their chemical interactions (metabolic pathways) and temporal concentration (flux) rates. Metabolomics is the third pillar (in combination with genomics and proteomics) of systems biology[1], and involves the measurement and analysis of metabolism in complex biological systems[2] in conjunction with their external environment. Changes in the metabolic profiles are the earliest cellular response to environmental or physiological changes such as toxin exposure or disease state, so metabolomics may be capable of, e.g., detecting and diagnosing a disease in an early stage. Yet metabolomics is the least developed component of systems biology. This is due in part to the difficulties associated with the various extraction procedures that are often used to determine the metabolic content. While this yields the highest possible spectral (chemical) resolution, it involves sample destruction and runs the risk of sample denaturation/decomposition, thus being difficult to compare directly with live sample measurements. Hence, metabolomics techniques are needed that can be applied on intact biological objects and provide minimal disturbance to cells and tissues.

   Non- or minimally invasive NMR is one of the main techniques that can be employed for metabolic studies, as the relatively small molecular weight and the relatively large mobility make many of them 'visible' with standard NMR. Therefore, in biochemical and biomedical research *in vitro* and *in vivo* NMR spectroscopy is increasingly used in cells, tissues, animals, and humans to use the presence and concentrations of the NMR-observable metabolites for, e.g., drug evaluations, studies of metabolic pathways, tumor phenotype characterization, diagnosis, and studies of cell proliferation, apoptosis, and necrosis[3-9].

A major problem associated with NMR spectroscopy in intact biological tissues is that relatively large resonance line widths are observed, often 1-2 orders of magnitude larger than the widths measured in liquids using established NMR techniques. This is especially a problem for ^{1}H NMR, which is the most widely employed nucleus because of its large NMR sensitivity, but has a relatively small chemical shift range. As usually many metabolites contribute to the NMR signal, the result is a spectrum with severely overlapping spectral lines, which seriously hampers a quantitative analysis of the spectra, and sometimes even makes it impossible to assign the spectral lines unambiguously.

In biological samples the main mechanisms for this broadening are the local magnetic field gradients arising from variations in the isotropic bulk magnetic susceptibility near boundaries of intra- and extra-cellular structures. Then chemically equivalent nuclei experience different local magnetic fields, depending on their spatial localization, giving rise to line broadening. In principle, this broadening can be averaged to zero by the technique of magic angle spinning (MAS), where the sample is rotated about an axis making an angle of 54°44' relative to the external magnetic field[10]. However, a problem is that in a standard MAS experiment spinning speeds of a kHz or more are required in order to avoid the occurrence of spinning sidebands (SSBs) in the spectra[11], which renders analysis of the spectra difficult again. At these spinning speeds the large centrifugal forces associated with this spinning cause severe structural damage in larger biological objects and sometimes even in the individual cells. Hence this method cannot be used to study metabolic processes in intact samples.

In solid state NMR several methods have been developed where slow MAS is combined with special radio frequency pulse sequences to eliminate spinning side bands or separate them from the isotropic spectrum so that a SSB-free high-resolution isotropic spectrum is obtained. It has been shown recently that two of these methods, phase-adjusted spinning sidebands (PASS) and phase-corrected magic angle turning (PHORMAT), can successfully be modified for applications in biological materials[12-19]. With PASS MAS speeds as low as 30 Hz can be employed, allowing *in vitro* studies of excised tissues and organs. With PHORMAT the MAS speed can be reduced to ~1 Hz, albeit at the cost of a reduced NMR sensitivity and a prolonged measuring time. Nevertheless, the ultra-low MAS speeds that can be employed make PHORMAT amenable for *in vivo* applications, as was demonstrated in a live mouse[16].

It is expected that *in vitro* and *in vivo* slow-MAS NMR can also play an increasingly important role in food science. NMR is already extensively applied in food science. Examples of such applications are food analysis[20], food processing[21], quality control[22], determining water and oil contents[23], and bacterial activities occurring during, e.g., food fermentation[24]. However, regarding the utility of metabolite spectroscopy in food science the same restrictions hold as mentioned above, unless the experiments are carried out on food stuff such as juices where the molecular mobility is so large that the spectral resolution becomes similar to that obtained with standard liquid-state NMR. Slow-MAS NMR opens the possibility to study food products where the spectral lines are broadened again. Examples of *in vitro* slow-MAS NMR applications are to study the maturation of fruit, meat, and other food products, the impact of storage conditions (duration, time, temperature) and preservatives on these maturation processes, the links between the metabolic profile and the food quality, and the effects of aerobic and anaerobic microbial activities on the food maturation and quality. Examples of applications of *in vivo* NMR spectroscopy include studies of links between the *in vivo* metabolic profiles in live species and their growth rates, yields, and, ultimately, the quality of the resulting food products, and the impact of the environment (light, nutrients, fodder, pesticides, genetic alterations) on the resulting food quality. For instance, it was found with ^{13}C NMR that the amounts of unsaturated and poly-unsaturated fatty acids in the adipose tissue of animals and humans depend on their diet[25], so the meat quality obtained from meat-producing animals can be manipulated and optimised even before slaughtering the animals. And similar experiments can be carried out in live fish and plants.

In this article the principles and limitations of PASS and PHORMAT will be briefly discussed and illustrated with spectra obtained on a variety of biological objects. Finally, on-going work to develop PASS and PHORMAT spectroscopic imaging and to improve the performance of both techniques will be discussed.

2 THE PASS AND PHORMAT EXPERIMENTS

Elaborate descriptions of PASS and PHORMAT have been published elsewhere[12-19], and here we shall restrict ourselves by a brief summary of the main features. The PASS and basic PHORMAT R.F. sequences are given in Figures 1a and 1b, respectively (in practice

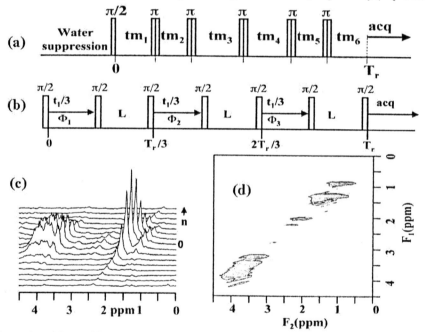

Figure 1   *(a): PASS R.F. pulse sequence; (b): The basic PHORMAT R.F. pulse sequence. Φ_1, Φ_2, and Φ_3 are precession angles, L is the storage period; (c): Stacked plot of water-suppressed 300 MHz ^1H 40 Hz PASS spectra of excised rat liver; (d); 2D water-suppressed 300 MHz ^1H 1 Hz PHORMAT spectrum of excised rat liver.*

a more complicated PHORMAT sequence is used, including a water suppression sequence prior to the last 90^0 pulse and additional 180^0 pulses to improve the base plane of the 2D spectrum[13, 16, 17]).

PASS is a constant-evolution-time 2D experiment with a duration equal to the rotor period T_r, during which five π pulses are applied with time intervals tm_1 to tm_6[26]. In PASS the center band spectrum and the SSB spectra are separated by order. This is achieved by acquiring the signal after a series of PASS experiments with n different values of the time intervals tm_1 to tm_6, where n denotes the total number of center band and side band spectra that has to be resolved. Then after 2D Fourier transformation a series of spectra is obtained that separates the center band (n=0) and side band spectra. Advantages of PASS are a good NMR sensitivity and a relatively short measuring time (seconds to minutes). A disadvantage is that the signal is observed after one rotor period, which means that the spinning speed has to be large compared to the intrinsic spin-spin relaxation rate $(T_2)^{-1}$ in order to avoid signal losses. For biological samples this means that spinning speeds of 30 Hz or larger have to be used. Figure 1c shows a stack of sixteen 300 MHz ^{1}H PASS spectra obtained on excised rat liver rotating at 40 Hz. Water suppression is obtained by preceding the PASS sequence by a water suppression sequence such as DANTE (Delays Alternating with Nutations for Tailored Excitation)[27].

PHORMAT is a regular 2D NMR experiment. The methodology of PHORMAT is based on the Magic Angle Hopping (MAH) experiment[28], where the sample is hopped over angles of $120^°$ about an axis at the magic angle. During each sample position the magnetization is brought in the transverse plane and allowed to dephase for a variable evolution time $t_1/3$. Then RF storage pulses are applied to orient the magnetization parallel to the external magnetic field B_0 during consecutive hopping periods. In PHORMAT the sample is spun slowly and continuously instead of hopped, and the effect of $120^°$ hopping is achieved by synchronizing the pulses to 1/3 of the rotor periods[14, 17, 29, 30]. This continuous rotation makes PHORMAT much easier to implement than MAH. Then with proper phase-cycling steps after 2D Fourier transformation a pure absorption-mode 2D spectrum is obtained where the isotropic and anisotropic information are separated. The main advantage of PHORMAT is that the spinning speed has to be large compared to the spin-lattice relaxation rate $(T_1)^{-1}$ of the spins rather than $(T_2)^{-1}$ in order to avoid signal attenuation. This means that in biological samples spinning speeds as low as 1 Hz can be employed. Disadvantages of PHORMAT are a reduced NMR sensitivity and relatively long measuring times (minutes to hours). Nevertheless, for *in vivo* applications PHORMAT is the only method that can be applied. Figure 1d shows the 2D spectrum of the excised rat liver, obtained with 300 MHz ^{1}H PHORMAT while spinning the sample at 1 Hz, illustrating the reduction in line width in the isotropic (F1) dimension. This is also illustrated in Figure 2, where the liver spectra obtained in a static sample and with 4 kHz MAS, the center band 40 Hz PASS spectrum, and the isotropic 1 Hz PHORMAT spectrum are shown. It follows that with all MAS techniques similar spectral resolution enhancements are obtained. The relative intensities of the various lines are somewhat different in the MAS spectra, which are due to variable T_1 and T_2 signal attenuations for the various lines in the PHORMAT and PASS experiment, respectively, associated with the used spinning speeds. Also, for the PASS spectrum only the center band spectrum is given. Measuring the T_1 and T_2 values of the metabolites, and shifting the PASS SSB spectra and adding them to the center band spectrum will correct these correct intensity distortions.

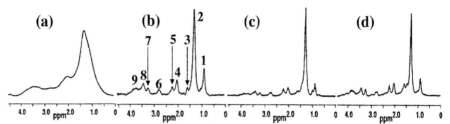

Figure 2   *300 MHz water-suppressed ^1H spectrum of excised rat liver obtained on a static sample (a) and the isotropic spectra obtained with 1 Hz PHORMAT (b), 40 Hz PASS (c), and 4 kHz standard MAS (d).   Line assignments: 1: triglycerides CH_3 terminal, or neutral amino acid methyl, valine, leucine, isoleucine methyl; 2: triglycerides $-(CH_2)_n$, and/or lactate methyl; 3: triglycerides OOC-CH$_2$-CH_2-; 4: triglycerides CH=CH-CH_2-CH$_2$; 5: triglycerides OOC-CH_2-CH$_2$-; 6: triglycerides CH=CH-CH_2-CH=CH; 7: total choline methyl, phosphocholine methyl, and β-glucose methine; 8, 9: glucose and glycogen methylene and methine.*

2 APPLICATIONS

In this section examples will be given of slow-MAS applications in a variety of biological objects. These objects are not all food products, but are chosen to illustrate the wide applicability of the slow-MAS methodology. All *in vitro* experiments have been carried out on a Chemagnetics Infinity spectrometer, equipped with a 7 Tesla magnet (corresponding to a proton Larmor frequency of 300 MHz). The samples were inserted into a 6 mm I.D. rotor. The *in vivo* measurements were performed using a Varian UnityPlus console and a wide-bore (30 cm) 2 T magnet, corresponding to a proton frequency of 86 MHz, and a rotor with an I.D. of 2 cm.

2.1 PASS in meat and seeds

Figure 3 shows static and PASS spectra of excised rabbit muscle (a) and two types of

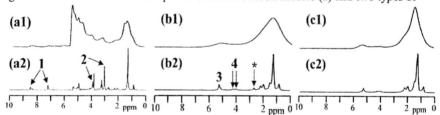

Figure 3   *300 MHz ^1H spectra of excised rabbit muscle (a), sesame seeds (b), and peanuts (c), obtained on a static sample (a1-c1) and with 40 Hz (a2) and 80 Hz (b2, c2) PASS. Line assignments: 1: histidine; 2: creatine; 3: glycerol and fatty acid methine; 4: glycerol methylene. The other lines are assigned in the caption of Figure 2.*

seeds, sesame seeds (b) and peanuts (c). The ^{1}H PASS experiments on the muscle were used in combination with ^{31}P NMR and ^{1}H water T_2-NMR to study post mortem changes in the water-holding capacity (WHC) and metabolic profiles in this tissue[31], processes of interest for the conversion of muscle into meat and the meat quality development. For this investigation the use of slow MAS was essential, as it made it possible to perform all the NMR experiments on a same sample while keeping the sample and its structure intact, which is a prerequisite for not affecting the WHC. The PASS experiments were carried out while spinning the rotor at 40 Hz. Hence the maximal centrifugal force of this 6 mm I.D. rotor was 39 times the gravitational force G. In a separate standard centrifuge experiment it was established that a 40G force did not damage the structural integrity of the sample. It can be observed from Figures 3a1 and 3a2 that another advantage of the slow-MAS approach is that the water suppression in the isotropic dimension is more effective. The reason is that in static samples the susceptibility gradients can broaden the water line to such and extent that the wings of the line are overlapping with adjacent metabolite intensities. Hence this water intensity cannot be suppressed without suppressing the metabolite lines as well, and a considerable residual water intensity is measured, cf. Figure 3a1. In PASS and PHORMAT the anisotropic broadening is eliminated in the isotropic dimension, resulting in a much better separation between the water and metabolite lines, as is evidenced in Figure 3a2. For the PASS experiments on the oily seeds no water suppression was needed, and the spectra are dominated by triacyl glycerides. The peanut and sesame spectra look very similar, except that in the sesame seed spectrum a line is observed at 2.7 ppm (the line marked with an asterisk in Figure 3b2) that is not measured in the peanuts. This line is the same as line 6 in Figure 2b, and originates from methylene peaks sandwiched between two ethylenic groups. Hence in the sesame seeds the amount of polyunsaturated fats is larger than that in the peanuts, in accordance with other findings[32,33]. This might indicate that the sesame seeds are healthier than peanuts.

2.2 PASS in cell systems

2.2.1 Xenopus Laevis oocytes. In addition to susceptibility gradients arising from inter-cellular structures, intra-cellular compartments can cause considerable susceptibility broadening as well. This is illustrated in Figure 4, where experimental results are shown obtained on oocytes from *Xenopus Laevis*, an African frog, which are being used

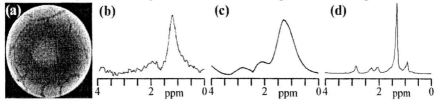

Figure 4   a): 500 MHz MR microscopic image of a 800 μm O.D. oocyte; b): 500 MHz ^{1}H spectrum of a 200x200x200 μm³ volume inside the cytoplasm; c) static 300 MHz ^{1}H spectrum of a large number of similar oocytes; d) 300 MHz ^{1}H spectrum of the same oocytes obtained with 40 Hz PASS.

extensively as model cells for stress studies. Figure 4a shows a 500 MHz MR micro-image of an excised single stage-IV oocyte with a diameter of 800 microns, showing the nucleus in the centre and some blood vessels surrounding the oocyte. Figure 4b shows a 500 MHz ^{1}H spectrum obtained in the cytoplasm. The spectrum is dominated by lipids and the spectral resolution is limited again, comparable to, e.g., the static liver spectrum, c.f. Figure 2a, where both intra- and intercellular susceptibility gradients are contributing to the observed line width. Figure 4c shows the 300 MHz ^{1}H spectrum obtained on a static sample of similar oocytes after inserting them in the rotor. The spectral resolution is less than the spectrum shown in Figure 4b, probably caused by a combination of the lower NMR frequency employed and additional susceptibility gradients induced by neighbouring cells and the rotor material. Figure 4d shows the spectrum obtained with 300 MHz ^{1}H PASS while spinning the sample at 40 Hz. The gain in spectral resolution is evident again, even compared with the spectrum obtained within a single cell and a higher frequency (Figure 4b), indicating that both intra- and intercellular susceptibility broadenings can be eliminated with slow MAS.

2.2.2 Microbial systems. Slow-MAS spectroscopy is also of importance for studies of microbial systems, especially when the cells are densely packed together or when they are attached to non-microbial surfaces. We investigated the impact of slow MAS in *Shewanella oneidensis* MR-1 microbes, dissimilatory metal-reducing bacteria that have the potential for application in the remediation of contaminated surface and ground waters. Figures 5a and 5b show the static and 40 Hz PASS metabolite spectra of MR-1, harvested

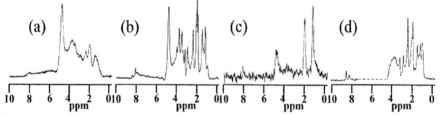

Figure 5   *Water-suppressed 300 MHz spectra of densely packed Shewanella oneidensis MR-1 microbes grown on agar plates. a): static spectrum of MR-1 grown for two days; b): 40 Hz PASS spectrum of MR-1 grown for two days. The line assignments are in part still under investigation; c): 40 Hz PASS spectrum of MR-1 grown for five days; d): 40 Hz PASS spectrum of MR-1, grown for five days and 2.5 hrs after administering pyruvate.*

from agar plates onto which the cells have been grown aerobically for 2 days. It follows that also for this densely packed cell system slow-MAS provides significantly more informative spectra. Spectrum 5c was obtained after growing the cells on agar plates for 5 days without adding nutrients. The cells basically shut down and only two remaining lines are observed at 1.3 ppm (lactate or methylene chains) and 2.0 ppm (acetate) (the third line at 4.7 ppm is residual water intensity). Figure 5d shows the revival of the metabolic activities when a nutrient is added. Here the PASS spectrum is shown obtained 2.5 hours after adding 20 mm^3 of a 200 mM solution of pyruvate to the cell system. Almost all the metabolite peaks observed in Figure 5b were regenerated, including succinate (2.4 ppm) and formate (8.5 ppm). These measurements illustrate that PASS can be used to follow metabolic activities in real time and under realistic conditions.

2.3 *In vivo* PHORMAT

So far we have performed all our in vivo experiments on mice, animals that are extensively used in biomedical and pharmaceutical research. We first investigated the impact of spinning a mouse in a magnetic field on its health, and have spun a mouse in a 2 T field up to 2 Hz for 70 minutes and up to 8 Hz for 40 minutes, without observing any apparent long-term health problems[16]. Figure 6A shows a picture of the mouse-MAS

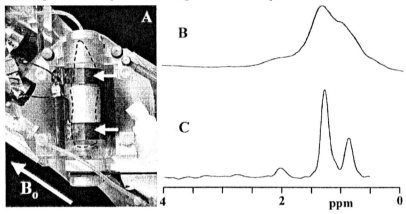

Figure 6 *A): A Top part of the mouse-MAS probe with a ~70 days old anaesthetized female BALBc mouse in the rotor. The dotted line indicates the mouse contour; B): 86 MHz ^1H spectrum of the part of the mouse body between the arrows in (A) obtained on a stationary mouse; C): 86 MHz ^1H PHORMAT spectrum of the same body area while spinning the animal at 1.5 Hz.*

probe, and Figures 6B and 6C show (water-suppressed) 86 MHz ^1H spectra obtained in a stationary mouse with a standard NMR experiment and with PHORMAT while subjecting the animal to 1.5 Hz MAS, respectively. The spectrum is dominated by the lipid signals arising mainly from the adipose tissue. It follows that even in this relatively low external field a significant enhancement in the spectral resolution is obtained, which makes it possible, e.g., to determine the intensities of the lines due to mono-unsaturated and polyunsaturated fatty acid chains (2.0 ppm and 2.0+2.7 ppm, respectively). In fact, the isotropic line widths are similar as those observed with PASS and PHORMAT in the 7T field, indicating that the resolution of the PHORMAT spectra increases almost linearly proportional with the external field.

3 CONCLUSION

It has been shown that slow-MAS NMR metabolite spectroscopy provides a significant enhancement in the resolution of proton NMR metabolite spectra in prokaryotic and eukaryotic cells, excised tissues, and live animals, thus significantly increasing the significance of NMR for metabolomic studies. Although several of the examples given above were not carried out on objects of direct interest to the food industry, it should be straight-forward to extend the use of the slow-MAS approach for food-science applications as well. Moreover, for in vivo studies it should be relatively easy to build MAS probes capable of spinning larger animals such as rabbits, provided that a magnet with a proper bore size is available. And even in these cases slow spinning may not cause harmful effects. For instance, the maximum centrifugal force inside a rotor with a radius of 10 cm, spinning at 1 Hz, is 0.4 times the gravitational force G, comparable to the value experienced in a rotor with a radius of 1 cm and spinning at ~ 3 Hz, a speed that a mouse inside this rotor survived without any apparent problems.

It has been mentioned already that the main advantage of slow-MAS spectroscopy is that intact living objects can be investigated. This is especially useful when the dynamics in cellular processes are studied, and several *in vitro* and *in vivo* applications of importance in food science have already been mentioned in the Introduction. It needs to be emphasized, however, that more work is needed to further improve the methodologies. For example, both PASS and PHORMAT need to be combined with pulsed-field-gradient NMR so that the spatial distribution of the metabolites can be determined rather than the overall spectra, the sensitivity of the PHORMAT needs to be enhanced, e.g., by implementing multiple spin-echo acquisition[15], and the PHORMAT measuring time needs to be reduced. These and other improvements are currently being undertaken in our laboratory. It is expected that if these improvements are successful, the slow-MAS approach could become a breakthrough in the utility of NMR spectroscopy in several bioscience disciplines, including food science.

Acknowledgments

The research was performed in the Environmental Molecular Sciences Laboratory (a national scientific user facility sponsored by the Department of Energy's Office of Biological and Environmental Research (BER)) located at Pacific Northwest National Laboratory, and operated for DOE by Battelle. The research on the rabbit muscles was supported by EMSL operational funding and the Danish Veterinary and Agricultural Research Council (SJVF). The microbial work was funded by PNNL's Biological Systems Initiative (BSI). The development of the slow-MAS methodologies and the other applications were supported by BSI, the Department of Energy Office of Biological and Environmental Research Program under Grant 22342 KP-14-02-01, and by the National Institute of Biomedical Imaging and Bioengineering under Grant R21 EB003293-01. The authors thank Eric Ackerman and Jeffrey McLean for providing the oocyte and microbial samples, respectively, Donald Rommereim for assistance in the animal care and handling, and Paul Majors and Kevin Minard for performing the MR microscopy work on the single *Xenopus* oocytes.

References

1. W. Weckwerth, *Ann. Rev. of Plant Biol.*, 2003, **54**:669.
2. J.C. Lindon, E. Holmes, M.E. Bollard, E.G. Stanley and J.K. Nicholson. *Biomarkers*, 2004, **9**,1.

3. R. Kreis, *J. Progr. in NMR Spectr.*, 1997, **31**, 155.
4. S.K. Mukherji and M. Castillo, eds. *Clinical Applications of Magnetic Resonance Spectroscopy.* John Wiley, New York, 1998.
5. N. Salibi N, and M.A. Brown, *Clinical MR spectroscopy*, John Wiley, New York, 1998.
6. F. Traber, W. Block, S. Flacke, H.H. Schild, and R. Lamerichs, *Medicamundi*, 2000, **44**, 12.
7. I.R. Young, ed., *Biomedical Magnetic Resonance Imaging and Spectroscopy*, Volumes I and II, John Wiley, New York, 2000.
8. M.R. Tosi, G. Fini, A. Reggiani, and V. Tugnoli, *Int. J. of Molec. Med.*, 2002, **9**, 299.
9. M.G. Swanson, D.B. Vigneron, Z.L. Tabatabai, R.G. Males, L. Schmitt, P.R. Carroll, J.K. James, R.E. Hurd and J. Kurhanewicz, *Magn. Reson. Med.*, 2003, **50**, 944.
10. E.R. Andrew and R.G. Eades, *Nature*, 1959, **183**, 1802.
11. P. Weybright, K. Millis, N. Campbell, D.G. Cory and S. Singer, *Magn Reson Med.*, 1998, **39**, 337.
12. R.A. Wind, J.Z. Hu, and D.N. Rommereim, *Magn. Reson. Med.*, 2001, **46**, 213.
13. J.Z. Hu, D.N. Rommereim, and R.A. Wind, *Magn. Reson. Med.*, 2002, **47**, 829.
14. J.Z. Hu, and R.A. Wind, *J. Magn. Reson.*, 2002, **159**, 92.
15. J.Z. Hu, and R.A. Wind, *J. Magn. Reson.*, 2003, **163**, 149.
16. R.A. Wind, J.Z. Hu, and D.N. Rommereim, *Magn. Reson. Med.*, 2003, **50**, 1113.
17. R.A. Wind and J.Z. Hu, Recent *Res. Devel. Magnetism & Magnetic Mat.*, 2003, **1**, 147.
18. R.A. Wind and J.Z. Hu, *US Pat.*, 2003, 6,653,832 and 6,670,811.
19. R.A. Wind and J.Z. Hu, *Encycl. of Anal. Science* , 2004 (in press).
20. E. Ibanez and A. Cifuentes, *Critical Rev. in Food Science and Nutrition*, 2001, **41**, 413.
21. L.F. Gladden, *Chem. Eng. Science*, 1994, **49**, 3339.
22. F. Capozzi, M.A. Cremomini, C. Luchinat, G. Placucci and C. Vignali, *J. Magn. Reson.*, 1999, **138**, 277.
23. G. Rubel, *J. Amer.Oil Chem. Soc.*, 1994, **71**, 1057.
24. A. Laws, Y.C. Gu and V. Marshall, *Biotechn. Advances*, 2001, **19**, 597.
25. E.L. Thomas and J.D. Bell, Body fat metabolism: Observation by MR imaging and spectroscopy, in *Biomedical Magnetic Resonance Imaging and Spectroscopy*, ed., I.R. Young, John Wiley, New York, 2000, pp.837-845.
26. O.N. Antzutkin, S.C. Shekar and M.H. Levitt, *J. Magn. Reson.,* 1995, **A115**, 7.
27. G.A. Morris and R. Freeman, *J. Magn. Reson.*, 1978, **29**, 433.
28. A. Bax, N.M. Szeverenyi and G.E. Maciel, *J. Magn. Reson.*, 1983, **52**, 147.
29. Z. Gan, *J. Am. Chem. Soc.*, 1992, **114**, 8307.
30. J.Z. Hu JZ, W. Wang, F. Liu, M.S. Solum, D.W. Alderman, R.J. Pugmire and D.M. Grant, *J. Magn. Reson.*, 1995, **A 113**, 210.
31. H.C. Bertram, J.Z. Hu, D.N. Rommereim , R. A. Wind, and H.J. Andersen, *J. Agric. Food Chem.*, 2004, **52**, 2681.
32. A.H. Bahkali, M.A. Hussain, A.Y. Basahy, *Int. J. of Food Sciences and Nutrition*, 1998, **49**, 409.
33. T.H. Sanders, *J. of Agricult. and Food Chem.*, 2001, **49**, 2349.

PROTON RELAXATION IN CRYSTALLINE AND GLASSY SUGARS

Peter Belton

School of Chemical Sciences and Pharmacy, University of East Anglia, Norwich NR4 7TJ, UK. Email: p.belton@uea.ac.uk

1 INTRODUCTION.

Simple sugars are important in food science both as the building blocks of complex polysaccharides and in their own right as nutrients and functional ingredients. Considerable effort has been put into the understanding of the dynamics of complex polysaccharides, for example in cell walls, but much less effort has been put into the understanding of the dynamics of simple sugars in sugars in the solid state. This seems to be a necessary requirement to understanding the dynamics of the more complex systems and of materials such as sugar glasses. Measurements were made of the relaxation times of a number of sugars over a range of temperatures. The data were treated by fitting the results to the well-known Kubo-Tomita equations[1].

2 THEORETICAL CONSIDERATIONS

Spin lattice relaxation in the laboratory frame of reference may be adequately modelled by the assumption of simple rotation of the relaxing entities[2]. Typically there will be more than one of these so that the relaxation rate, R_1 is given by:

$$R_1 = \frac{1}{T_1} = \sum_{i \geq 1} C_i \left[\frac{\tau_{ci}}{1 + \omega_0^2 \tau_{ci}^2} + \frac{4\tau_{ci}}{1 + 4\omega_0^2 \tau_{ci}^2} \right] \quad (1)$$

where ω_0 is the proton Larmor frequency and τ_c the rotational correlation time of the motions responsible for the spin lattice relaxation, which are assumed to follow the Arrhenius activation law

$$\tau_{ci} = \tau_{0i} \exp\left(\frac{E_{ai}}{RT}\right) \quad (2)$$

where τ_0 is the pre-exponential factor corresponding to the rotational correlation time at infinite temperature, E_a the activation energy and R the gas constant.
C in the equation 1 is the relaxation constant. For methyl groups undergoing rapid three-fold rotation, C has a relationship with that part of the van Vleck second moment (ΔM_2) modulated by the motions given in the first part of equation 3:

$$C = \frac{2}{3}\Delta M_2 \gamma^2 = \frac{27}{20N} \cdot \frac{\gamma^4 \hbar^2}{r^6} \qquad (3)$$

N, is the total number of protons relaxed by the methyl group rotation and mean inter-proton distance in the methyl group, r, as given in second part of equation 3. γ is the proton magnetogyric ratio. Other forms of equation 3 may be derived for other relaxing entities. For spin lattice relaxation in the rotating frame a similar equation applies

$$R_{1\rho} = \frac{1}{T_{1\rho}} = \frac{3}{2}\sum_{i\geq 1} C_i \left[\frac{\tau_{ci}}{1+4\omega_{ei}^2 \tau_{ci}^2} + \frac{5}{3}\times\frac{\tau_{ci}}{1+\omega_o^2 \tau_{ci}^2} + \frac{2}{3}\times\frac{\tau_{ci}}{1+4\omega_o^2 \tau_{ci}^2}\right] \qquad (4)$$

where ω_0, τ_c and C have the same meaning as in equation 1. ω_e is the effective field for relaxation in frequency units. This is generated by the radio frequency field. ω_0 is generated by the main magnetic field, thus the effective frequencies for relaxation in the laboratory frame are hundreds or tens of megahertz and those for the rotating frame are tens of kilohertz. This means that each of these measurements is sensitive to different ranges of rotational correlation time.

The form of the equations 1 and 4 imply that at some value of the product $\omega\tau$ a maximum in relaxation rate will be reached. This is readily observable but for relaxation in the rotating frame it occurs for values of τ of the order of microseconds and for the laboratory frame for values of the order of picoseconds. Clearly quite different regimes of motion may be studied by each measurement.

The free induction decay is a measure of the transverse relaxation rate. Typically for pure solids below the melting point or glass transition the free induction decay can be approximated by the Abragam expression[3]:

$$I(t) = I_G \exp\left(\frac{-a^2 t^2}{2}\right) * \frac{\sin bt}{bt} \qquad (5)$$

The residual rigid lattice second moment is defined as $M_{2r} = a^2 + b^2/3$. In general, M_{2r} will decrease in value when the motions of some of the protons contributing to the local field has a correlation time, τ_c, which is comparable to the inverse of line-width,

i.e., $\qquad \tau_c \leq (\gamma \sqrt{M_{2r}})^{-1} \qquad (6)$

It will also be affected by the interproton distances, generally as temperature increases so will these, as such it will be expected that the second moment will show gradual decrease with increasing temperature. Superimposed on this will be sharper changes as motions of different groups meet the criterion given in equation 6.

3 RESULTS

The spin lattice relaxation behaviour of fucose and methyl-fucose in the laboratory frame exemplify the effects of methyl group relaxation[4], both compounds contain methyl groups but methyl fucose contains an additional O-methyl. The results are shown in Figure 1.

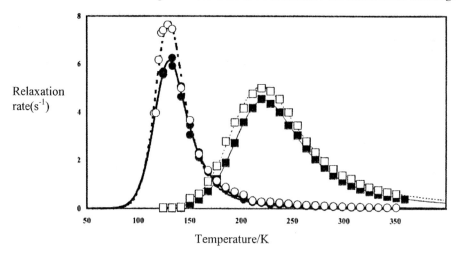

Figure 1   *Spin-lattice relaxation in the rotating frame of fucose (squares) and methyl fucose (circles). Open symbols represent deuterium exchanged samples. Lines are fits to the data.*

These results, which are fairly typical for sugars, demonstrate the maxima arising from the rotations of the methyl groups, O-methyl relaxation maxima occurring at lower temperatures than the methyl relaxation maxima. The rise in the rates when exchangeable protons are replaced with deuterium at first sight appears anomalous. However, since methyl rotation is the principle source of relaxation, the methyl groups must relax all the protons in the compound. Exchange of these protons reduces the number of protons to be relaxed. This corresponds to a decrease in N in equation 3 which corresponds to an increase in R_1 in equation 1. In effect the relaxation load of the methyl groups is reduced and hence relaxation is faster.

If the rotating frame relaxation rate is measured, the methyl maxima will occur at lower temperatures and other motions may be observed. In Figure 2 the example is given of crystalline rhamnose[2]. The methyl maximum occurs at a lower temperature than could be achieved on the spectrometer but the rise in relaxation rate below 150K indicates where it would be observed. At high temperatures another smaller maximum is observed which disappears on deuterium exchange. This must arise from exchangeable protons and is therefore due to water or hydroxyl groups. Calculations of the value of C (equations 3 and 4) are consistent with the values calculated for water.

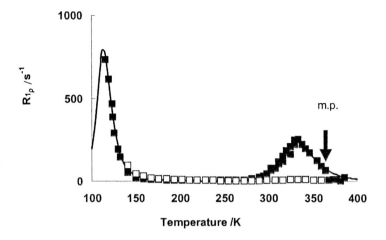

Figure 2   *Spin-lattice relaxation in the rotating frame of crystalline rhamnose*

When the rigid lattice second moments calculated from the free induction decays using Abragam equation are plotted against temperature (Figure 3) there is a general decline in the value between 130 and 300K and a much sharper drop at about 300K.

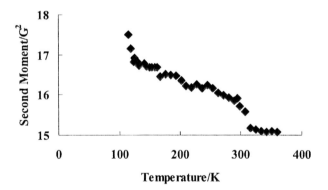

Figure 3   *Variation in rigid lattice second moment of crystalline rhamnose with temperature*

The value of the sharp change is about $0.7G^2$ which is close to that calculated for water[2], thus confirming the conclusion.

In the case of glassy sugars similar behaviour is observed, but there are some subtle differences. The $R_{1\rho}$ data for rhamnose[2] (figure 4) show a large maximum at around 300 K that can only be explained as resulting from the motion of hydroxyl groups and motions of the whole molecule[2]. In addition a second motion causing a gradual increase in the value of $R_{1\rho}$ can be seen to start at about 170 K; this disappears on deuterium exchange and is due to water.

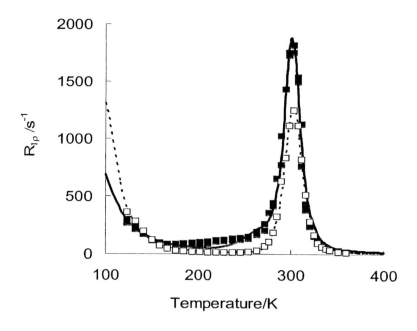

Figure 4 *Spin-lattice relaxation in the rotating frame of rhamnose glass*

Relaxation in the laboratory frame shows similar behaviour although the high temperature maxima occur around 350 K. If the relaxation processes in both frames are due to the same motion then it should be possible to predict the rotating frame relaxation from the laboratory frame motion and vice versa. Figure 5 shows an attempt to do this. Clearly the rotating frame data fail to predict either the position of the laboratory frame maximum or its intensity. The reasons for this can be seen by considering the values for activation energy, E_a and the relaxation constant, C. for the rotating frame process these are: 114 kJ/mol and 17.8×10^8 rads^2/s^2 respectively and for the laboratory frame process they are 67 kJ/mol and 27.09 rads^2/s^2. The conclusion must be that within the different temperature ranges covered by the rotating frame and laboratory frame maxima two different processes occur. The rotating frame process, which occurs close to the glass transition, is a restricted motion requiring a large activation energy. This motion transforms, at higher temperatures, to one which is less restricted and has lower activation energy. There is thus a transition that occurs around 350K. Using all the data it is possible to construct a map of the changes in correlation time with temperature. This is done for a hypothetical methyl rhamnose hydrate glass in Figure 6. It illustrates that most of the motions occurring in the frequency range of 10^{12} to 10^4 Hz remain unaffected by the glass transition and that a transition in motion occurs at a point between the glass transition and the melting point. This implies that the nature of the liquid formed after the glass transition is not in the same state as that formed after the melting point.

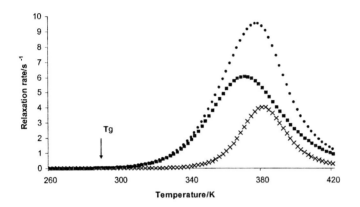

Figure 5 *Observed and predicted values of R_1 for methyl rhamnose; squares: observed data, crosses: predicted from $R_{1\rho}$ data, circles: combined rates*

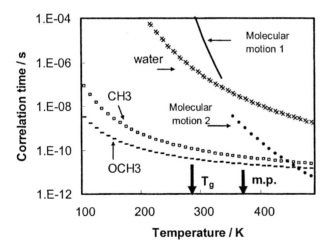

Figure 6 *Correlation time map for a hypothetical methyl rhamnose hydrate glass*

References

1   R. Kubo and K. Tomita, *J.Phys.Soc.Jpn.,* 1954, *9*, 888

2   H.R. Tang Y.L. Wang, and P.S. Belton, *Phys. Chem. Chem. Phys.,* 2004, **6**, 3694

3   A. Abragam, *The Principles of Nuclear Magnetism,* Oxford University Press, Oxford, 1961. p 120

4   Y. L. Wang, H. R. Tang and P. S Belton, *J.Phys.Chem. B,* 2002, **106**, 12834

New NMR Methods and Instrumentation

TOWARDS ON-LINE NMR SENSORS

B.P. Hills and K.M. Wright

Institute of Food Research, Norwich Research Park, Colney, Norwich, NR4 7UA, UK

1 INTRODUCTION

Since its discovery the phenomenon of NMR has been widely exploited for clinical diagnosis in hospitals and as a research tool in analytical laboratories throughout the world. However, apart from relatively minor off-line quality control applications it has so far failed to be developed as a sensor in the industrial sector. This is surprising because almost every other type of spectroscopy, from Near Infra Red to X-rays, has been exploited as an on-line sensor[1]. There are at least five reasons usually cited for this failure, namely
a) MRI is too expensive because it requires expensive gradient and RF amplifiers and controllers and is usually based around super-conducting magnet technology.
b) MRI is too delicate for a factory environment because it requires high magnetic field homogeneity and signal/noise is generally low.
c) MRI is too slow for industrial conveyors moving at typical speeds of 2m/s because the available acquisition time for moving samples is reduced to just a few tens of milliseconds.
d) There is insufficient time to polarise most food samples moving at conveyor velocities, because samples need to travel a distance $5T_1v$ for complete polarisation. With a conveyor velocity, v, of 2m/s and a typical T_1 of 1 second this amounts to ten meters, which is impractical.
e) MRI requires highly skilled (and expensive) operators and technical support.
These are valid criticisms, but they are based on current NMR protocols. In this paper we show how these objections can be overcome by abandoning conventional NMR and exploiting a hitherto neglected aspect of NMR, namely "motional relativity". As we shall see in the next section, this uses the movement of the sample through constant magnetic fields to create an NMR signal.

2 THE MOTIONAL RELATIVITY PRINCIPLE IN NMR

2.1 Motional relativity and radiofrequency excitation

Most commercial NMR spectrometers use pulsed radiofrequency fields to excite transverse magnetisation in a stationary, polarised sample. The effect of an on-resonance RF pulse of time-dependent amplitude, $B_1(t)$, is to tip the longitudinal magnetization vector in the sample through an angle, θ, in the rotating frame, such that

$$\theta = \int dt \, \gamma B_1(t) \tag{1}$$

The alternative way of achieving the same result, which exploits motional relativity, is to set up a localised region of space containing a spatially characterized, time-invariant, continuous radiofrequency field, $B_1(x,y,z)$ which is transverse to the main magnetic field, B_0. The excitation is then achieved by motion of the sample through this localised volume of RF radiation. The tip angle, θ, in the on-resonance rotating frame for a spin moving in one dimension (along the z axis) through an on-resonance RF field, $B_1(z)$, is again given by the integral,

$$\theta = \int dt \, \gamma B_1(t) = \int dz \, (dt/dz) \, \gamma B_1(z) \tag{2}$$

For a single spin entering the region with initial velocity v and acceleration, a,

$$z = vt + at^2/2 \tag{3}$$

so that, time can be replaced by distance as a variable such that,

$$\theta = \int dz \, \gamma B_1(z) / \, [v^2 + 2az]^{1/2} \tag{4}$$

There are major advantages with this alternative excitation protocol in on-line situations. First, because the RF irradiation is continuous there is no need for expensive RF pulse programmers. Second, rapid translation of the sample through the RF field at velocities of several meters per second no longer requires extended regions of radiofrequency B_1 homogeneity along the direction of motion. Even quite inhomogeneous radiofrequency fields are effective because only the spatial integral of the RF field enters equation (4). Thirdly, because the RF field is no longer pulsed there are no induced Eddy currents or associated coil ring-down phenomena. Equation (4) also shows that there is no reason to be restricted to samples moving at constant velocities. Accelerating samples, such as powders falling under gravity through the RF coil, can be excited in the same way.

   This excitation mode, which we call the "SM" mode (for Sample Motion) is especially suited to simple on-line applications requiring only one or two excitation pulses. As we shall show, on-line multi-pulse sequences are best implemented with conventional pulsed RF methods.

2.2 Motional relativity and the NMR signal

The FID of the spins emerging from a region of radiofrequency excitation will differ from the FID of a pulsed RF experiment because spins at different locations in a finite-sized sample emerge from the RF field at different times and therefore at time zero, taken as the time the whole sample has emerged from the RF excitation region, have different phases due to free precession in the main field, B_0. Neglecting relaxation, the signal dS from a volume element (a slice) located at z at time t in the laboratory frame is

$$dS(z,t) = \rho(z).\exp[i\varphi(z,t)]dz \tag{5}$$

where $\rho(z)$ is the local slice transverse magnetisation density and $\varphi(z,t)$ is the phase. In the laboratory frame, $\varphi(z,t) = (\varphi_0 + \omega_0 t)$ where the initial phase distribution, $\varphi_0(z)$, is, neglecting acceleration, simply $\omega_0(z/v)$. The total signal is therefore

$$S(t) = \exp(i\omega_0 t) \int dz \rho(z) \exp[i\omega_0(z/v)] \qquad (6)$$

This is a Fourier transform with a wavevector, $k_0 = \omega_0/v$, which should be contrasted with the conventional FID on a stationary sample in a pulsed RF experiment, which is $S(k_0=0,t)$. The emergence of the Fourier transform suggests that image profiling is possible if we impose a constant, time-invariant field gradient, G, oriented along the direction of sample motion. An analogous derivation shows that in this case,

$$\varphi(z,t) = \varphi_0 + \omega_0 t + \gamma Gzt + \tfrac{1}{2}\gamma Gvt^2 \qquad (7)$$

and the Fourier transform relationship is now,

$$S(t) = \exp[i\omega_0 t].\ \exp[i\ \tfrac{1}{2}\gamma Gvt^2].\ \int_w dz \rho(z) \exp[i(k_0 + k)z].\ \exp[i\ \gamma(G/v)z^2] \qquad (8)$$

Here the wavevector k is defined in the usual way as γGt. Providing $\gamma G/v$ is very small (the weak gradient, high velocity approximation) we can neglect the term in z^2 which prevents inverse Fourier transformation. The pre-exponential factor $\exp[i\omega_0 t]$ can also be set to unity if we perform the experiment "on-resonance", in which case

$$S(t) = \exp[i\ \tfrac{1}{2}\gamma Gvt^2].\ \int dz \rho(z) \exp[i(k_0 + k)z] \qquad (9)$$

Equation (9) shows that the image, $\rho(z)$, is obtained as the inverse Fourier transform of $S(t)\exp[-i\tfrac{1}{2}\gamma Gvt^2]$. Note that in this one-dimensional imaging experiment all fields are time-invariant and only the sample moves. This derivation can be generalised to include spin-echoes, arbitrary tip angles, accelerating samples and non-linear field gradients and provides the basis for novel on-line imaging protocols. Pulsed RF excitation of a sample moving through a constant z-gradient corresponds to the $k_0 = 0$ limit in equation (9).

The motional relativity principle is trivially simple for gradients transverse to the direction of sample motion because the vector product **G.v** is then zero so the signal is not motionally modulated. Passage through a spatially localised time-invariant transverse gradient is therefore exactly equivalent to applying a pulsed transverse gradient to a stationary sample.

2.3 Motional relativity and coil design

To exploit motional relativity and maximize signal acquisition times with samples moving at several meters per second all RF, B_0, shim and gradient coils have had to be redesigned as cylindrical modules that can be extended to arbitrary lengths along the direction of

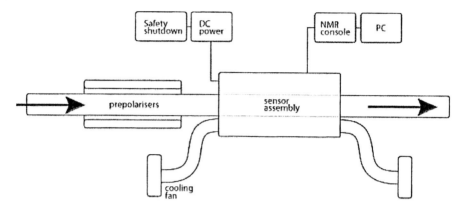

Figure 1 *Schematic of the low cost, on-line NMR sensor prototype based on an air-cooled solenoid electromagnet and permanent magnet prepolarisers*

sample motion. For this reason, and to minimise costs, a prototype sensor has been built which uses an air-cooled solenoid electromagnet (see Figure 1). The solenoid creates a low B_0-field of up to about 0.06T (proton frequency 2.55MHz), limited by the need for efficient air cooling. The B_0 field can be shimmed to order z^3 with specially designed concentric, cylindrical z and z^2 shim coils, which reduce the linewidth of a water phantom to about 150Hz. To explore a range of multipulse sequences the prototype sensor uses conventional pulsed RF excitation rather than the SM mode. Designing a suitable RF coil with cylindrical symmetry was not, however, straightforward. A simple solenoid RF coil cannot be used because B_1 needs to be transverse to B_0. Saddle and birdcage coils give the desired transverse field but have limited B_1 homogeneity and long pulse lengths. To circumvent these difficulties a series-wound, Eddy-compensated, tilted solenoid RF coil was designed. This contains solenoid-like coils tilted at 45^0 and uses the induced currents in a set of concentric vertical inner coils to remove the undesired longitudinal B_1 field component. The coil can be extended to arbitrary lengths limited only by the need to avoid destructive interference when the wire length approaches half the RF wavelength. The decreasing filling factor also becomes a problem with extended coils. For this reason separate transmitter and receiver coils are desirable, and are essential for the SM excitation mode.

2.4 Prepolarisation and the adiabatic principle

The discussion in sections 2.1 and 2.2 tacitly assumed that the sample is fully polarised (i.e. contains its equilibrium longitudinal magnetisation) before entering the RF excitation coil. This, however, is an unreasonable assumption with many food samples characterised by long T_1's. A sample with a T_1 of one second travelling at 2 meters/s would need to travel 10 meters ($5T_1v$) before it is fully polarised, which implies an unreasonably long solenoid magnet. To circumvent this difficulty the prototype sensor was equipped with a prepolarising module (see Figure 1). This contains low-cost rectangular ferrite permanent magnets creating a higher field (ca. 0.12T) than that in the solenoid and oriented transverse to the sample motion. Samples first pass through the prepolarising module before entering the main solenoid sensor field. The higher prepolarising field means that sufficiently high polarisation in the lower solenoid field can be achieved in a shorter distance. In fact samples with short T_1 will be superpolarised relative to the solenoid field.

Two other aspects of prepolarisation are noteworthy. The sample emerges from the prepolariser with longitudinal magnetisation oriented transverse to the direction of motion; whereas the solenoid magnet creates a B_0 field oriented parallel to the direction of motion. The sample polarisation therefore needs to be rotated through 90^0 before NMR is possible. However simple mechanical rotation will not suffice because the sample necessarily travels through a region of highly inhomogeneous stray field in the gap between the prepolariser and the solenoid. On first consideration this would appear to prevent any coherent on-line NMR. Fortunately, however, the adiabatic principle[2] rescues the situation and ensures that, in most circumstances, the magnetisation in each spin isochromat in the sample follows the changing orientation of its local external field, so that the magnetisation is realigned inside the solenoid magnet and NMR is still possible. For the same reason the sample can be mechanically rotated as it travels through the prepolariser to ensure a more uniform polarisation.

Of course, the adiabatic principle merely determines the rate of reorientation of the longitudinal magnetisation not its magnitude, which is determined by longitudinal relaxation. It is therefore noteworthy that longitudinal relaxation as samples travel through the prepolariser, gap and solenoid magnet can be exploited to give a new type of image contrast, namely the degree of polarisation, $P(\mathbf{r})$. This depends not only on the field-dependent longitudinal relaxation time, $T_1(\gamma B_0(z))$ but also on the changing asymptotic longitudinal magnetisation, $\chi B_0(z)$. The stray field in the gap can correspond to low resonance frequencies, $\gamma B_0(z)$, of just a few tens of kHz so that $T_1(\gamma B_0(z))$ in the gap differs significantly from its usual high-field value and needs to be measured with a field-cycling spectrometer. Polarisation contrast can be varied in many ways, either by altering the length of the stray field gap or by changing the prepolariser length or even the prepolarising field strength by varying the separation, thickness and type of ferrite or neodynium-ferrite block magnets used in the prepolariser.

2.5 Single-shot on-line pulse sequences

Because each sample travels through the sensor only once there is no possibility of using conventional pulse sequences based on repeated acquisition and/or phase cycling. Instead maximum information must be extracted in a single passage through the sensor. This is not a problem with measurements of transverse relaxation because the Hahn-echo and CPMG sequences can be operated in a single-shot mode without phase cycling or repeated acquisition. Moreover each spin echo acquired in a fixed gradient can be processed using the protocol of section 2.2 to give image profiles. If more than one RF coil is located along the sample path then transverse magnetisation can be stored as longitudinal magnetisation for passage between RF coils with a single 90_{-x} flip-back pulse. Alternatively, if signal/noise permits, smaller tip angles can be used in each RF coil, thereby reserving longitudinal magnetisation for use down-stream. More significant problems arise when sample quality correlates with T_1 because conventional sequences such as saturation or inversion recovery require repeated acquisition. Fortunately the image can be T_1-weighted either by using polarisation contrast or by varying the degree of saturation by very rapid repeat acquisition on a timescale short compared to T_1 and the sample residence time in the RF coil. For more direct single-shot on-line measurement of T_1 we have developed a novel

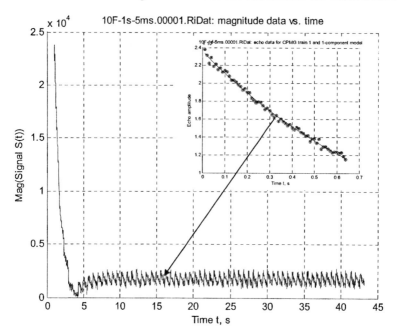

Figure 2 *Representative data for a 15% sucrose solution acquired with the FIRE sequence for single-shot, on-line T_1 and T_2 measurement. The saw-tooth structure arises from truncated CPMG echo trains, one of which is expanded and fitted in the inset.*

pulse sequence called FIRE (for Fast Inversion REcovery). This single-shot sequence actually measures both T_1 and T_2 and is similar to an inversion recovery sequence except that the signal is acquired not with an FID but with a truncated CPMG sequence of n-echoes. A 90_{-x} flip-back pulse on the $(n+1)$th echo of each truncated CPMG train then stores longitudinal magnetisation for a delay time t_1 before the next truncated CPMG sequence. The magnetisation eventually reaches a steady-state value that depends on both T_1 and T_2 so that fitting each CPMG echo train and the amplitude of the first echo of each truncated CPMG sequence gives both T_1 and T_2. In multicompartment or multicomponent systems the method gives the component relaxation times and the proton fractions. Figure 2 shows an example of the FIRE sequence.

3 OPTIMISING ON-LINE ACQUISITION PROTOCOLS

The on-line acquisition protocols outlined in the previous section allow measurement of NMR parameters such as T_1's, T_2's, self-diffusion coefficients and proton densities on moving samples. However, to be useful it is first necessary to know which combination of these NMR parameters best correlates with the sample quality factor of interest. In some

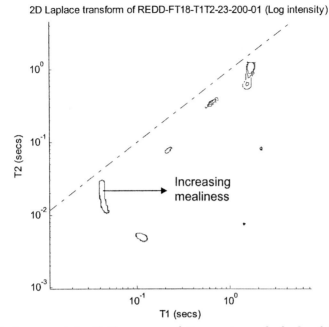

2D Laplace transform of REDD-FT18-T1T2-23-200-01 (Log intensity)

Figure 3 *A representative T_1-T_2 cross correlation spectrum of a fresh red delicious apple. Arrows indicate the shift to longer T_1 with increasing mealiness.*

applications, such as measurements of solid/liquid ratios or oil or water contents, the well-established off-line method can simply be transferred to the on-line situation. However there are many, more subtle, quality factors where it is unclear which, if any, combination of NMR parameter correlates with the quality defect. In such cases it is helpful to first explore the NMR parameter space with off-line multidimensional methods. To illustrate this strategy consider the problem of the on-line detection of mealiness in apples. The first step is to compare the two-dimensional T_1-T_2 correlation spectra[3-5] of healthy and mealy apples. A representative spectrum is reproduced in Figure 3. Such spectra show whether longitudinal or transverse relaxation is the best probe of mealiness and whether short or long relaxation time components should be measured. In fact the data show that the short T_1 of peaks assigned to cell wall water become longer in mealy apples which suggests that the on-line FIRE sequence might be able to discriminate degrees of apple mealiness, though this has yet to be tested. Earlier off-line measurements at a higher field of 100MHz showed that mealiness shortens the T_2^* of the main vacuolar water peak[6] but this correlation is not observed at the low frequency used in the on-line system, confirming that it is a susceptibility-induced dephasing effect caused by increasing numbers of air spaces in the tissue.

The relationship between polarisation contrast, $P(\mathbf{r})$, and sample quality is of particular interest. T_1 at the low frequencies associated with the stray field in the prepolariser-solenoid gap will, in general, be much smaller than that usually measured in the tens of MHz frequency range. To optimise polarisation contrast for quality determination it is therefore best to measure the whole T_1 frequency dispersion for both high and low quality samples with an off-line field-cycling spectrometer, though, to date, we are unaware of any such studies.

4 POTENTIAL APPLICATIONS OF ON-LINE NMR SENSORS

The commercial potential for low-cost NMR sensors in the food manufacturing sector is enormous. Many NMR papers have shown the existence of useful correlations between NMR parameters and food quality factors and these can, potentially, be developed into on-line quality control protocols. However the example of apple mealiness discussed in the previous section emphasises the need to test these correlations at lower spectrometer frequencies than those usually used in commercial spectrometers.

Space does not permit a comprehensive review of potential applications of the on-line technology. NMR studies of quality defects in horticultural products (fruit and vegetables) have been reviewed recently[7] and reveals numerous potential applications including the detection of bruising, infection and physiological defects induced by over-ripening or long-term storage. Foreign body detection in processed food is another area ripe for exploitation. Bone fragments in processed meat and fish as well as metal, glass and plastic foreign bodies introduced during manufacture are a constant concern in the food industry.

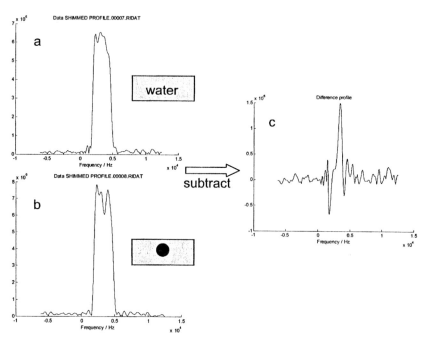

Figure 4 *One-dimensional image profiles acquired with the prototype on-line sensor.*
a) A pure water sample b) water with a foreign body (rubber bung)
c) The difference profile (a-b) highlighting the presence of the foreign body

Existing dual X-ray sensors are expensive and are only able to detect foreign bodies containing atoms with high X-ray scattering cross-sections such as the calcium in bone fragments. Plastic objects, insects, wooden splinters and other organic materials like cigarette ends therefore pass undetected through X-ray scanners. This, in principle, is not a limitation with an on-line MRI foreign body detector, though low signal/noise might limit detection sensitivity. Figure 4 illustrates the feasibility of such on-line MRI foreign body detection. Figure 4a shows the profile of a bottle of pure water measured in the prototype sensor. Figure 4b corresponds to the same sample except that a rubber bung has been introduced. Figure 4c shows the difference profile and therefore highlights the presence of the foreign body.

Besides foreign body detection there are a host of quality concerns related to the oil, fat and water content of meat, dairy, egg and cereal commodities. This includes spoilage changes such as creaming and undesired phase separations. Process control is another area where NMR sensors could make a significant impact. In process control the spatial aspect of imaging is not really needed and an on-line NMR sensor could continuously and non-invasively monitor the composition and/or viscosity and/or temperature of continuously extruded or pumped food material. Such monitoring relies on the dependence of the NMR relaxation times and signal amplitudes on the processing variables.

5 FUTURE DEVELOPMENTS

The prototype sensor has been designed to work with only a single RF coil, but in a commercial version it would be desirable to incorporate several RF coils positioned at various points along the solenoid magnet. This would permit increased sample throughput and/or multiple NMR measurements on each sample. Of course, care must then be taken to correctly synchronise and gate the transition and receiver modes of the coils. As a first step to developing such multiplexed on-line acquisition, computer simulations of the effect of repositioning the RF coil in the solenoid magnet of the prototype sensor were undertaken, focusing on changes to the CPMG echo train acquired from samples moving through off-centre RF coils.

Figure 5a shows the B_0, B_1 and sample profiles used in the simulation and Figure 5b shows the simulated CPMG echo decay envelopes for samples moving at the indicated speeds through an RF coil located one-third of the way down the solenoid magnet. It can be seen that even when the RF coil is well off-centre, it is still possible to determine meaningful T_2's out to acquisition times of about 40ms with samples moving at 1.25m/s.

Two- and three-dimensional on-line image acquisition is another area ripe for future development. To date only one-dimensional projection profiles have been acquired from samples moving through the prototype sensor. However 2-dimensional back-projection imaging is possible using specially designed transverse gradient coils extending down the direction of motion. Such coils have been designed for both pulsed and continuous operation. Unfortunately these further aspects have yet to be patent-protected so cannot be discussed in detail here.

Safety and automation are other critical issues that need serious attention in any attempt at commercialisation. Commercial NMR sensors need to operate in fully automated mode, be immune to the frequent cleansing operations so vital in the food manufacturing sector and be capable of shutting down safely in the event of overheating, power failure and the like. The prototype has sensors continuously monitoring the power supplies to the magnet and cooling fans, the magnet temperature and the flow of the air through the magnet. If abnormal readings are reported the whole system is automatically shut down in programmed steps designed to avoid high back-voltages that could wreck the electronics.

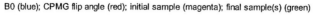

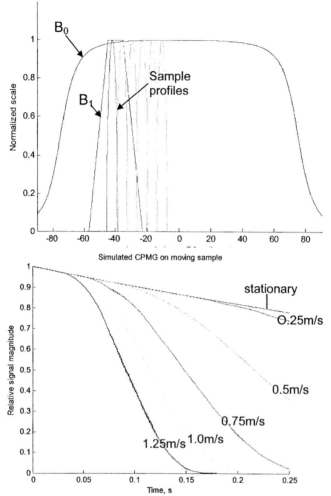

Figure 5. *A) This shows the trapezoid B_1 of an off-centre RF coil, the B_0 map of the solenoid magnet and the semi-ellipsoidal profile of a spherical sample used in the simulation. B) shows the calculated CPMG echo decay envelopes for the indicated sample velocities.*

6 CONCLUSIONS

Motional relativity presents an alternative way of performing NMR that holds great promise for the eventual development of commercial, low-cost on-line NMR sensors. Despite the encouraging results emerging from the prototype sensor many hurdles remain to be overcome. While the feasibility of on-line foreign body detection has been established, the detection limits for different classes of foreign body have yet to be investigated and this aspect is critical to the sensors commercial viability. Similarly the robustness of the NMR-quality correlations on moving samples in low fields remains to be researched with statistically significant numbers of samples.

At a more fundamental level, many aspects of the underlying spin-physics of motional relativity remain to be explored. For example, the SM excitation mode can be applied to fluids flowing coherently through a pipe. In the SM mode the NMR signal will be automatically modulated by the fluid velocity field and therefore, potentially, contains useful rheological information. A theoretical framework for this has been outlined but much remains to be done before the motional relativity principle can be exploited for on-line rheological measurements.

Acknowledgments

The authors wish to thank Niusa Marigheto for measuring the T_1-T_2 correlation spectra of fresh and mealy apples. The BBSRC is thanked for a grant under the SBRI scheme and a Bridge-Link award from DEFRA is gratefully acknowledged.

References

1. M.Edwards (editor), *Detecting Foreign Bodies in Food*, Woodhead Publishing Ltd., Cambridge, 2004.
2. C.P.Slichter, *Principles of Magnetic Resonance*, Springer-Verlag, Berlin and Heidelberg GmbH & Co., 1989, chapter 2.
3. B.P.Hills, S.Benamira, N.Marigheto and K.M.Wright, *Applied Magnetic Resonance* (in press).
4. Song, Y.-Q., Venkataramanan, L., Hurlimann, M.D., Flaum, M., Frulla, P., and Straley, C., *Journal of Magnetic Resonance* 2002,**154**,:261-268,
5. Godefroy, S., Creamer, L.K., Watkinson, P.J., and Callaghan, P.T., page 85 in Magnetic Resonance in Food Science, Ed Belton.P.S., Gil, A.M., Webb, G.A. and Rutledge, D., The Royal Society of Chemistry, Cambridge, 2002.
6. P.Barreiro, A.Moya, E.Correa, M.Ruiz-Altisent, M.Fernandez-Valle, A.Peirs, K.M.Wright and B.P.Hills, Applied Magnetic Resonance, 2002, **22**, 387.
7. B.P Hills and C.J. Clark, Annual reports on NMR spectroscopy, 2003, **50**, 76.

DETERMINATION OF THE APPARENT DIFFUSION COEFFICIENT OF WATER IN RED BLOOD CELL BY HIGH FIELD PFG-NMR USING VARIOUS PULSE SEQUENCES

Y.S. Hong, [1] S.O.Ro, [2] H.K Lee,[3] V.I.Volkov, [4] and C.H.Lee [1]

[1] Graduated School of Biotechnology, Korea University, Seoul, Korea.
  E-mail: chlee@korea.ac.kr
[2] Dept. of Nursing, Shin Heung College. Eujungbu, Kyunggido, Korea
[3] Korea Occupational Medicine Institute, Eujungbu, Kyunggido, Korea
[4] Karpov Institute of Physical Chemistry, Moscow, Russia

1 INTRODUCTION

The material transport through living cell wall is an important physiological function for the maintenance of life and regulation of metabolic processes. Water is the carrier of the moving materials such as nutrients, enzymes and other metabolites in the organism. The osmotic movement of water itself regulates the ionic strength between compartments in the tissue and keeps the equilibrium of the body for the homeostasis. Therefore, the rate of water transport through the cell wall is an important parameter for the cellular response on external stresses and metabolic conditions such as food and drug intake, pathological conditions as well as the inherited types of body constitution

In this respect, water permeability of red blood cell can be used for the evaluation of the physiological function of foods, drug response, diagnosis of physiological disorders and diseases, and also genetic characterization of body constitution

1.1 NMR techniques for measurement of water transport in biological cells

In the cell there are two known pathways of water transport, namely the phospholipid bilayer diffusion and the active transport through water channel proteins. The protein mediated transport covers around half of the total water movement as measured by the blockers like *p*-chloromercuri benzoate (*p*-CMB).[1,2]

Nuclear Magnetic Resonance technique is the most suitable method of water diffusion investigation of biological cells. There are two methods most widely investigated for this purpose; NMR relaxation technique and Pulsed field gradient (PFG) NMR techniques.

1.1.1 NMR relaxation technique. In the relaxation paramagnetic doping method, the water exchange time, τ_e, is calculated from the longitudinal (T_1) and transverse (T_2) ^{1}H relaxation times of the water in the interior and exterior compartment by Mn^{2+}-doping ^{1}H NMR.

$$\frac{1}{\tau_e} = \frac{1}{T_{2a'}} - \frac{1}{T_{2i}} \cdots \cdots (1)$$

The membrane permeability of water diffusion is calculated from the rate constant, k, and the ratio of volume and surface area of cell (V/S).

$$P^d = k\frac{V}{S} \quad \text{where} \quad k = \frac{1}{\tau_e} \cdots \cdots (2)$$

By using this method Benga et al.[3-5] determined the diffusional water permeability and the activation energy for water diffusion of red blood cell of different species of animal. The water permeability and the activation energy varied with the animal species from 4.6-7.4×10^{-5}m/s, and 22-45 KJ/mol, respectively.

1.1.2 Pulsed field gradient NMR technique. In PFG NMR technique, stimulated spin-echo sequence (STE) and longitudinal eddy current delay sequence (LED) are widely used. Figure 1 shows the stimulated echo pulse sequences with the magnetic field gradient pulses. The attenuation curves of cellular waters show non-exponential decays. For the multiphase system the following equation is used.

$$A(g)/A(0) = p_i \sum_{1}^{n} \exp(-\gamma^2 \cdot g^2 \cdot \delta^2 \cdot D_{si} \cdot t_d) \cdots \cdots (3)$$

where A(g) is the spin echo amplitude at gradient strength of g, A(0) is the spin echo amplitude at g=0, γ is the gyromagnetic ratio, $t_d (= \Delta - \delta/3)$ is the diffusion time, and D_{si} is the self-diffusion coefficient of i-th component, and p_i is the relative amounts of nuclei (proton in water) belong to the molecules. By using the t_d dependence of self-diffusion coefficient, the permeability, P^d, and the restricted area (cell size), a, are estimated by the following relationship for pore system.

$$\frac{1}{D_p} = \frac{1}{D_0} + \frac{1}{P^{eff} \cdot a} \cdots \cdots (4)$$

where D_0 is non-restricted self-diffusion coefficient at $t_d \to 0$, D_p is hindered self-diffusion coefficient. By using scaling approach, the calculated dependencies $D_s^{eff}(t_d)$ proportional to t_d^{-1} are analyzed.

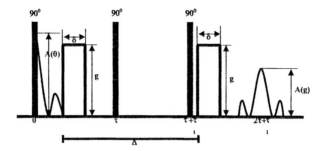

Figure 1   *Stimulated echo pulse sequence with the magnetic field gradient pulses. Here τ is the time interval between the first and second RF pulses, and τ_l is the time interval between the second and the third ones. Δ is the interval between the gradient pulses, δ is duration of the magnetic field gradient pulses, and g is the amplitude of the gradient pulse. The gradient pulse is rectangular and oriented along the Z axis.*

$$D_s^{\mathit{eff}}(t_d) = \frac{D_0[D_s(t_d) - D_p]}{D_0 - D_s(t_d)} \dots \dots (5)$$

From $D_s^{\mathit{eff}}(t_d)$ slope, which is proportional to t_d^{-1}, the restricted size, a, is determined according to the Einstein equation ($a = (6 \cdot D_s^{\mathit{eff}} \cdot t_d)^{1/2}$). The permeability, P^d, is calculated from D_p value according to equation (4). [6-9]

In order to calculate water molecules' residence (exchange) time, τ_e, in the cell and the exchange rate constant, k, two or three phases model consisting slow exponential decays are used.

$$A(g)/A(0) = p_i \cdot \exp(-\gamma^2 \cdot \delta^2 \cdot g^2 \cdot t_d \cdot D_{si}) + p_e \cdot \exp(-\gamma^2 \cdot \delta^2 \cdot g^2 \cdot t_d \cdot D_{se}) \dots \dots (6)$$

where p_i and D_{si} is the apparent intracellular water amount and self-diffusion coefficient, and p_e and D_{se} is the apparent extracellular water amount and self-diffusion coefficient, respectively. Exchange time is calculated from dependence of apparent intracellular water population or amount p_i on diffusion time t_d, $p_i(t_d)$, which is approximated by exponential function, $p_i = p(0) \cdot \exp(-\tau_e/t_d)$. [6-9]

The diffusional permeability, P^{eff}, is estimated from the same exchange rate constant as in equation (2).

1.2 Application in cellular water transport studies

Sorokina[10] demonstrated the three components water movements in red blood cell by using stimulated spin-echo sequence analysis of PFG-NMR. The three components comprised hydrated water in channels and lateral lipids (D_{sm}), intracellular water (D_{si}) and extracellular water (D_{se}). They found that the water exchange time of red blood cell of hypertensive patient was relatively insensitive of pH change from 6.55-7.4, compared to that of normal people. Latour et al.[2] reported that within 1 hr after the occlusion of the middle cerebral artery in animal models of focal ischemia, the effective (long-time) diffusion coefficient, D^{eff}, in the affected region of the brain decreased by up to 50%.

We have measured the water permeability of packed yeast and chlorella cells in water by stimulated spin-echo sequence in a home-built PGF-NMR[8,9]. The three component water behaviours were observed and they were assumed to be the intra- and extra-cellular waters exhibiting restricted diffusion, and the free water. The NMR frequency for protons was 63 MHz, the maximum value of the field gradient amplitude and the duration of the magnetic field gradient pulses were 50 T/m and 5×10^{-3} s, respectively. It allowed the measurement of the diffusion coefficient of water in yeast and chlorella from 10^{-9} to 10^{-15} m^2/s, and to observe diffusional decay change in three-order of magnitude. The latter is very important when this decay is non-exponential.

1.3 Objectives of present study

In our earlier study with red blood cell for the water permeability of smokers and non-smokers, the attenuation decay signal scattered different from yeast and chlorella cells, and we were not able to obtain reliable measurement of self-diffusion coefficient and permeability of water of red blood cell[7]. In order to overcome this problem, we studied the effect of pulse sequence on the water permeability measurement of red blood cell by using a 600 MHz PFG-NMR.

2 METHOD AND MATERIALS

2,1 Sample preparation

Blood samples were obtained by venipuncture from a healthy volunteer, centrifuged at 3000 rpm for 10 min. The cells were washed by resuspending three times in a phosphate buffer saline solution (pH 7.4). The final hematocrit was 82%. The packed red blood cell was filled in 5-mm (external diameter) NMR tube (series 300, Aldrich Chemcial, USA).

2.2 PFG-NMR measurements

Pulsed Field Gradient (PFG) NMR measurements were performed on a Bruker DMX600 spectrometer (proton frequency of 600.05 MHz) with a 5-mm triple resonance probe incorporating triple-axis gradient coils. For each diffusion experiment, measurements were conducted at 32 different values of gradient strength varying from 0.02 to 1.15 T/m with gradient pulse sequences based on a stimulated echo sequences. The gradient pulse sequences studied were stimulated echo (STE), bipolar pair stimulated echo (BPPSTE), longitudinal eddy current (LED), and bipolar pair longitudinal eddy current sequence (BPPLED), using sine-shaped bipolar gradient pulses and homospoil gradient pulses (Fig. 2). A longitudinal eddy current delay (Te) of 5 ms was used in LED and BPPLED sequence. Diffusion times $(t_d, t_d = \Delta - \delta / 3)$ were varied from 5ms to 100ms. At every gradient strength, eight scans were acquired with a recycle delay of 3 s. The temperature was 309.5 K. The data were analyzed by using the equations (3)-(6).

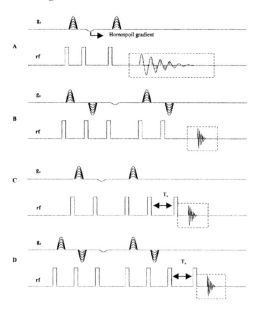

Figure 2 *600MHz PFG-NMR pulse sequences used for diffusion measurements, with radio frequency (rf) pulses of 90° represented by white bars and gradient pulses (g_z) by gray sinball-shaped bars: (A) pulsed-field gradient stimulated echo (STE) sequence, (B) STE sequence with bipolar gradient pairs (BPPSTE), (C) Longitudinal eddy current (LED) sequence, (D) LED sequence with bipolar gradient pairs (BPPLED). The homospoil gradient was applied all sequences. T_e represents the delay time before acquisition of final signal.*

3   RESULTS AND DISCUSSION

Figure 3 compares the stacked spectrum and their diffusion decay of water in human red blood cell of applying the stimulated echo sequence (STE) and bipolar pairs stimulated echo sequence (BPPSTE) in Fig. 2. The diffusion decay using STE sequence showed collapsed curve from any phase errors of spectrum compare to that using BPPSTE sequence. Figure 4 also shows phase distortion with longitudinal eddy current (LED) but best spectrum and diffusion decay were obtained by bipolar pairs longitudinal eddy current (BPPLED) sequence.

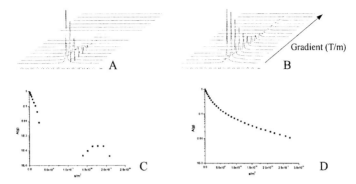

Figure 3 *600MHz ^1H PFG diffusion stacked plots and decays dependent on the bipolar of gradient pulses on STE. (A) and (B) represent the stacked plots using STE sequences without and with bipolar gradient pulses, respectively. (C) and (D) show the diffusion decays corresponding the (A) and (B), respectively.*

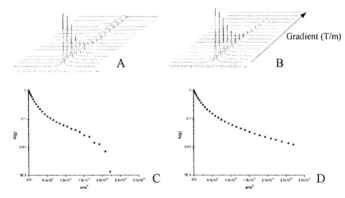

Figure 4 *600MHz ^1H PFG diffusion stacked plots and decays dependent on the bipolar of gradient pulses on LED sequences. (A) and (B) represent the stacked plots through LED sequences without and with bipolar gradient pulses, respectively. (C) and (D) show the diffusion decays corresponding the (A) and (B), respectively.*

The best diffusion decay without phase errors are obtained from BPPSTE and BPPLED sequence where matched pairs of field gradient pulses of opposite sign are used. The stacked spectrum and diffusion decay obtained with these two sequences are very similar. Since the BPPSTE sequence showed the phase distortion of spectrum with increasing gradient strength, the BPPLED sequence was preferred for high resolution PFG NMR spectrometer systems in measurement of diffusion coefficient of water in red blood cell (Fig. 5). The BPPSTE was more preferred option in high resolution diffusion-ordered spectroscopy (HR-DOSY) of quinine than BPPLED sequence because the BPPLED required longer phase cycling and gave slightly reduced sensitivity.[11] Our diffusion decays with 4- and 16-fold phase cycle were the same (data not shown). The phase distortion with BPPSTE was started to appear at 0.24 T/m (24 G/m) of gradient strength in this study (Fig. 5), whereas Pelta et. al has compared the diffusion spectrum at low gradient strength of 0.01 T/m (1 G/cm) so that they couldn't find phase and line shape errors with BPPSTE sequence. Therefore, the BPPLED sequence is more advantages sequence than BPPSTE due to reducing eddy current generation using relaxation delay time at high gradient amplitude.

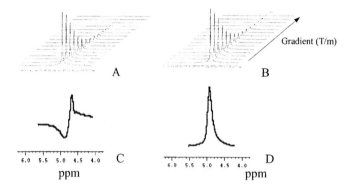

Figure 5   *The expanding water spectrum attenuated at gradient strength of 0.6 T/m during PFG measurements using BPPSTE (C) and BPPLED (D) sequences. A and B were extracted stacked plot from Fig. 3 and Fig. 4 using BPPSTE (A) and BPPLED (B), respectively. There was phase distortion in the water spectrum as the gradient strength was increased, even though STE sequence was incorporated to bipolar gradient due to eddy currents.*

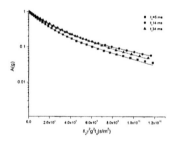

Figure 6   *The diffusion decay of water molecules on diffusion times in red blood cell by 600MHz PFG-NMR using bipolar longtudinal eddy current sequence (BPPLED). The maximum gradient strength was 1.15 T/m, respectively. Solid lines is the biexponential fitting curves according to Eq. (1).*

Figure 6 shows the diffusion decay curve of red blood cell by using BPPLED sequence in 600 MHz PFG-NMR. The non-restricted self-diffusion coefficient, D_0 , at t_d → 0, and hindered self-diffusion coefficient, D_p , were determined from the decay curve, and by using scaling approach, the calculated dependencies $D_s^{eff}(t_d)$ proportional to t_d^{-1} were analyzed (Figure 7). Table 1 shows the values of intracellular restriction size a , exchange time, τ , and permeabilities P^d and p^{eff} of the red blood cell of 6 volunteer men and women.

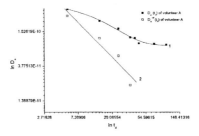

Figure 7   *The dependences of intracellular water self-diffusion coefficients D_s on diffusion time t_d in volunteer A. Curve 1 and 2 show the original and effective intracellular water self-diffusion coefficients.*

Table 1   *Apparent diffusion coefficient (ADC), average cell size a and permeabilities p^d, p^{eff} of red blood cell of volunteer men and women.*

Volunteers	ADC (m^2/s) **	a, μm	τ, ms	p^{eff}, m/s	p^d, m/s
A (F)	2.03	3.02	35	$2.22 \cdot 10^{-5}$	$4 \cdot 10^{-5}$
B (F)	2.02	2.94	30	$2.00 \cdot 10^{-5}$	$3.3 \cdot 10^{-5}$
C (M)*	1.68	2.00	30	$3.40 \cdot 10^{-5}$	$3.3 \cdot 10^{-5}$
D (M)	2.04	3.11	23	$1.86 \cdot 10^{-5}$	$4.3 \cdot 10^{-5}$
E (F)	1.94	1.98	36	$5.15 \cdot 10^{-5}$	$3.3 \cdot 10^{-5}$
F (M)*	1.61	1.93	31	$3.47 \cdot 10^{-5}$	$3.2 \cdot 10^{-5}$

F: female, M: male, * smoker, ** ADC: diffusion coefficient of water at 5 ms of diffusion time.

The apparent diffusion coefficient varied from 1.6 to $2.0 \cdot 10^{-10}$ m^2/s, The average cell size varied from 2 to 3 μm, and the permeability, p^{eff}, from $1.9 \cdot 10^{-5}$ to $3.1 \cdot 10^{-5}$ m^2/s. These values are in good agreement with the permeability, P^d, obtained by using the exchange rates of intracellular water molecules k and cellular volume to surface ratio V/S in Eq. (2). We were not able to conclude any differences between genders, smokers and non-smokers, and blood cholesterol level with this small number of sample.

So, we can see that scaling approach using D_0, D_p and $D_s^{eff}(t_d)$ used in in this study of red blood cell have given the same results with two compartment exchange model. Compared to exchange model, scaling approach allows us to estimate average cell size a and free intracellular water self-diffusion coefficient D_0. The value D_0 was about $1 \cdot 10^{-9}$ m^2/s that is three times less than bulk water self-diffusion coefficient at this experimental temperature of 36.5°C, which is $3 \cdot 10^{-9}$ m^2/s. It means that translational mobility of intercellular water molecules in RBC is three times less compare to bulk water.

4 CONCLUSION

As conclusion, the water permeability of red blood cell can be measured by PFG NMR techniques with proper pulse gradient sequence. Applying bipolar pair to the sequence reduced the phase distortion and improved the diffusion decay spectrum. The best spectrum and diffusion decay were obtained by bipolar pairs longitudinal eddy current (BPPLED) sequence. Scaling approach allows us to estimate average cell size a and free intracellular water self-diffusion coefficient D_0. Further studies are needed in order to evaluate the effect of food intake on the permeability of red blood cell by PFG-NMR technique.

Acknowledgements

This work was supported by grants from KOSEF (Korea Science and Engineering Foundation, # R01-2003-000-10717-0) and was done with V. I. Volkov invited by KISTEP (Korea Institute of Science & Technology Evaluation and Planning).

References

1 A. Finkelstein, 1987, Water Movement Through Lipid Bilayers, pores, and plasma membranes ,Wiley-Interscience, New York.
2 L L. Latour, K. Svoboda, P. P. Mitra and C. H. Sotak. Time-dependent diffusion of water in a biological model, *Proc. Natl. Acad. Sci. USA*, 1994, **91**, 1229.
3 G. Benga, S.M. Grieve, B.E. Chapman, C.H. Gallagher and P.W. Kuchel. *Comparative Hematology International*, 1999, **9**, 43.
4 G. Benga, P.W. Kuchel, B.E. Chapman, G.C. Cox, I. Ghiran and C.H. Gallagher, *Comparative Hematology International*, 2000, **10**, 1.
5 G. Benga, B.E. Chapman, G.C. Cox and P. W. Kuchel. *Cell Biology International*, 2003, **27**, 921-928
6 C.H. Lee, Y.S. Hong, V.D. Skirda and V.I. Volkov. *Magnetic Resonance in Food Science*, The Royal Society of Chemistry, Cambridge, UK, 2003, pp. 199-205
7 C.Y.J. Lee, H.W. Park, J.H. Song, K.C. Kim and C.H. Lee. 2002, *6th International Conference on Application of Magnetic Resonance in Food Science*, 4-6 September 2002, INA P-G, Paris
8. C.H. Cho, Y.S. Hong, K. Kang, V.I. Volkov, V. Skirda, C.Y.J. Lee and C.H. Lee. *Magnetic Resonance Imaging*, 2003, **21**, 1009.
9 K.J. Suh, Y.S. Hong, V.D. Skirda, V.I. Volkov, C.Y. Lee and C.H. Lee. *Biophysical Chemistry*, 2003, **104**, 121.
10 N. Sorokina, 2000, *PH.D. Thesis*, Kazan State University, Dept. of Molecular Physics, Russia
11 M.D. Pelta, H. Barjat, G.A. Morris, A.L. Davis and S.J. Hammond. *Magnetic Resonance in Chemistry*, 1998, **36**, 706

A NEW PRINCIPLE FOR UNIQUE SPECTRAL DECOMPOSITION OF 2D NMR DATA

Rasmus Bro[1], Peter Ibsen Hansen[1], Nanna Viereck[1], Marianne Dyrby[2], Henrik Toft Pedersen[3] and Søren B. Engelsen[1]

Dept. of Food Science, Quality & Technology, The Royal Veterinary and Agricultural University, 1958 Frederiksberg C, Denmark
[2] Umetrics AB, 211 34 Malmø, Sweden
[3] Virology & Molecular Toxicology, Novo Nordisk A/S, 2760 Måløv, Denmark

1 INTRODUCTION

Resolving pure component spectra from complex bilinear 2D NMR spectra of mixtures is highly desirable, but not possible at present. Extracting individual-analyte information from data such as diffusion-weighted spectra (DOSY - diffusion edited spectroscopy) would be important in many applications. The mathematical principle behind a new method for reaching this goal is outlined and its successful use is demonstrated on a series of mixtures in a severely reduced design. The potential of the new method applied to multi-parametric NMR spectroscopy appears to be inexhaustible and includes aiding tools in structural assignments and the recovery of pure spectra of impurities in low-rank mixtures.

1.1 **Multi-way data**

Multi-way analysis[20] is an emerging data analysis technique that has been successfully applied in several chemical fields[1, 2, 9, 10, 16, 18, 21]. The key issue in *multi-way* analysis is to have access to boxes of data rather than tables of data. Usually, a spectrum is measured for each sample. Data for several samples are then gathered in a matrix/table. If, for example, *a set* of spectra is obtained from a sample, then the data from just one sample are contained in a matrix. For several samples a box of data is obtained (Figure 1). Such *multi-way* data can be modelled with specialized tools that take particular advantage of the data format. Most notably, the so-called PARAFAC model is an interesting alternative to traditional data analysis tools, because it allows resolving complex mixture measurements into the underlying single-component spectra.

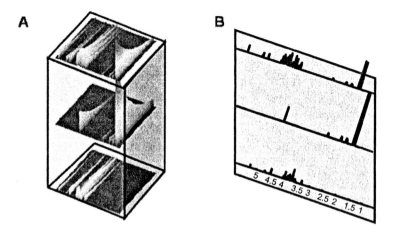

Figure 1   *Structure of multi-way data (A) and of ordinary two-way data tables (B). Rather than having one spectrum for each sample, a set of spectra can lead to multi-way data.*

1.1   Trilinear PARAFAC modeling

The PARAFAC model was developed in 1970 by R. Harshman[11]. The model provides a rational way to analyse multi-way data. The box of three-way data, called $\underline{X}$, is modelled as a sum of F contributions. Ideally, each of the F contributions describes one underlying chemical component. Each component consist of one score vector, $\mathbf{a}_f$, and two loading vectors, $\mathbf{b}_f$ and $\mathbf{c}_f$. If $\underline{X}$ is an $I{\times}J{\times}K$ array consisting of I samples measured at J variables at K occasions, then each score $\mathbf{a}_f$ will be a vector of length I and similar for $\mathbf{b}_f$ and $\mathbf{c}_f$. In scalar notation, the model can be written in terms of the individual elements of $\underline{X}$:

$$x_{ijk} = \sum_{f=1}^{F} a_{if}b_{jf}c_{kf}, \quad i=1,..,I; j=1,..J; k=1,..K \tag{1}$$

where a_{if} is the i'th element of $\mathbf{a}_f$ etc. Details of the model and its derivation can be found in the literature. The one feature of this model which distinguishes it markedly from, for example, principal component analysis (PCA)[13, 15, 17] is that no artificial constraints such as orthogonality need to be imposed to obtain a unique solution. If the fundamental laws prescribe that the data can be approximated as the above trilinear model, then PARAFAC can uniquely estimate these physical parameters. With PCA it is generally not possible to determine physically meaningful parameters, because there are infinitely many solutions to the same algebraic form as the PCA model. One "arbitrary" of these solutions is chosen in PCA on the basis of the constraints of orthogonality and maximum variance. The implication of the uniqueness properties of the PARAFAC model is that PARAFAC is able to resolve the underlying individual contributions directly from a mixture. Hence, PARAFAC performs *mathematical chromatography*.

2 EXAMPLE: HIGH-RESOLUTION NMR

In 2D diffusion-edited NMR the signal intensity is recorded as a function of chemical shift as well as of gradient strength:

$$I_{\delta gk} = \sum_{f=1}^{F} S_{\delta f} \cdot A_{gf} \cdot C_{kf} \tag{2}$$

where $S_{\delta f}$ denotes the spectral intensity at chemical shift δ for compound f, and where $A_{gf} = \exp\left[-D_f \gamma^2 g^2 \Delta' - R_f\right]$ denotes the attenuation due to diffusion (D_f) at gradient strength g for compound f, and C_{kf} denotes the concentration in sample k of compound f. When comparing the above model with the PARAFAC model, it follows that 2D diffusion-edited data can be separated into individual constituent spectra ($\mathbf{b}_f$), diffusion profiles ($\mathbf{c}_f$) and concentrations ($\mathbf{a}_f$). To date, such a mathematical separation has not been possible. An example will be used to show the results of applying PARAFAC modelling on 2D diffusion-edited data.

2.1 Experimental setup

2.1.1 Samples Mixtures of glucose, lactose and iso-leucine were made according to a reduced design (Figure 2). Only seven samples were used in this work. These samples were prepared with the three compounds at 0, 10 and 20 mM. The compounds were dissolved in a pH 6.0 phosphate buffer (0.1 M) in D_2O. The samples were prepared as 500 µl of the above described solutions with 50 µl 1mg/ml TSP (3-trimethylsilyl-1-[2,2,3,3-2H4] propionate) in D_2O added as chemical shift reference (0.0 ppm).

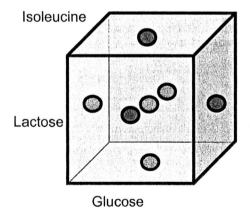

Figure 2 *Design of the three compounds (glucose, lactose and iso-leucine) at three concentration levels*

2.1.2 ^1H nuclear magnetic resonance. Proton spectra were measured at 298 K on a Bruker DRX600 operating at 14.1 T using a flow-NMR system and a 120 μl flow-probe. The pulse program used was a stimulated echo experiment with bipolar gradient pulses and a longitudinal eddy current delay as well as pre-saturation to suppress residual H_2O signal. The gradient was varied in 32 steps from 0.05 to 0.95 of maximum gradient power using a squared ramp.

2.2 Results

As can be seen in the subset of data shown in Figure 3, there are severe overlaps between the spectra and between the squared gradient attenuation profiles (from now on denoted diffusion profiles) of different analytes. Even though some spectral regions have limited overlap, the diffusion profiles are highly correlated throughout and hence, it is not possible to separate the mixture data into pure contributions immediately.

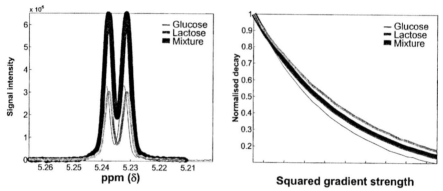

Figure 3   *Spectral region (left) and diffusion profiles (right) of the α-anomer proton of lactose, glucose and a mixture of the two*

When modelling the mixture data with a PARAFAC model, no assumptions on spectral band shape or distribution of concentration needs to be made. The only assumptions required for PARAFAC modelling are:

1. The data is low-rank trilinear
2. The number of components is decided by the user
3. The concentrations, spectra and profiles of different analytes are not identical

Based on the number of components and the measured data *only*, the PARAFAC solution is determined.

2.2.1 Elaboration on assumptions. It is perhaps instructive to consider the above verbal description of assumptions in somewhat more detail. The data has to be low-rank trilinear. This implies, for example, that the spectrum of an analyte does not change shape (significantly) at different gradient strengths. This is the crucial premise of the usefulness of the PARAFAC model. The model provides *one* spectrum for each analyte, so this can only be meaningful if indeed; the shape remains unaltered at different conditions. Experimental conditions have to be set so that this assumption is reasonably valid. This means proper tuning and matching as well as working in range of gradient strength where the output of the gradient amplifier is well defined for the selected diffusion time. Furthermore, the combination of diffusion time and maximum gradient power should not be set so signals becomes extinct, and it is important that the sample temperature is kept sufficiently low and constant in order to avoid convection current which will strongly influence the acquired signal. In addition, pH has to be constant from sample to sample to further avoid chemical shift differences.

Even though the low-rank trilinear requirement is not fulfilled, it is possible to perform meaningful multi-way analysis on this kind of data, but pure spectra etc. will not be obtained. Examples of such data analysis have been investigated by Dyrby *et al.* when exploring lipoprotein signals in DOSY spectra from human blood serum [5, 6].

The number of components has to be decided by the user. For the given data, it is known that the number of components should be three: one for each analyte. In general, however, such results are not known and the number of components therefore has to be determined from the data. This is similar to the situation in principal component analysis[7] and there are numerous tools for deciding on the appropriate number of components in PARAFAC[3, 4, 20]. Many of these are even simpler than for principal component analysis, because the intrinsic uniqueness of the PARAFAC model can help in guiding[4, 12].

The final assumption mentioned above is that the profiles are not identical. This unique condition can be stated more mathematically and stringently[14, 19], but conceptually it suffices to know that the main issue is that if, for example, the two self-diffusion coefficients are identical within the signal-to-noise ratio, then complete uniqueness cannot be guaranteed. One remarkable property that follows directly from the more stringent uniqueness results is that, in principle, only two samples are needed, even when many more than two underlying analytes are modelled. This is used in so-called second-order calibration[8, 22] where one sample contains analyte and unknown interferents and the other sample is standard with known non-zero concentration of the analyte. From two such samples the concentration of the analyte can be determined in the unknown sample.

2.3 The PARAFAC model

The so-called scores of the PARAFAC model contain estimates of the concentrations of the three analytes. The scores are shown in Figure 4.

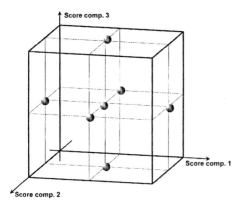

Figure 4   *Estimated relative concentrations from PARAFAC model of the mixture data*

When comparing with Figure 2 it is remarkable to see how well PARAFAC resolves the concentrations. However, since there is no way to determine the concentration in specific molar units from a spectrum, the PARAFAC model cannot provide concentration estimates on an absolute level. Hence, the scores shown in Figure 4 are relative concentrations that need to be scaled to the concentration of at least one sample or alternatively to a unit-concentration spectrum. This lack of absolute level is not specifically related to PARAFAC, but holds for *all* indirect measurements.

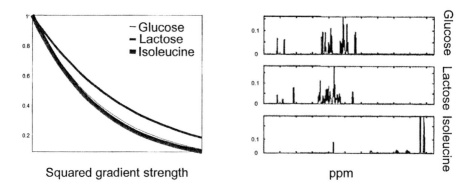

Figure 5   *The PARAFAC resolved diffusion profiles (left) and spectra (right)*

In Figure 5, the estimated gradient profiles and spectra are shown. In order to verify the quality of these estimates, the residuals are shown in Figure 6 comparing the estimated spectra with measured pure spectra. The similarity between estimated and measured real spectra is evident.

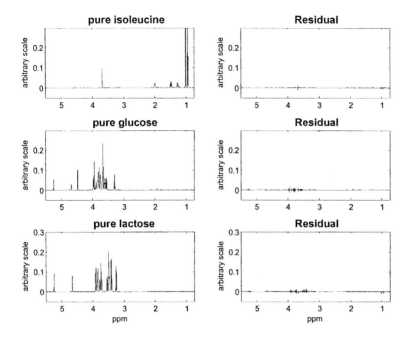

Figure 6 *Pure spectra and residuals between pure measured spectra and PARAFAC estimates of spectra*

3 CONCLUSION

The results presented show that diffusion-ordered NMR spectroscopy data fulfil the low-rank trilinearity requirement, since it is possible to extract the pure component spectra by application of a PARAFAC model. This is accomplished, despite the fact that the selected chemical analytes show a relatively high degree of overlapping NMR spectra and have relatively similar self-diffusion coefficients. Furthermore, it is accomplished with a limited number of samples, all including mixtures of the three chemical components. It is important to stress that the latter is not a requirement, in the sense that if one or more samples did not contain all three components, the analysis would still be handled equally well. These samples would simply show zero concentration for the missing component(s).

By measuring one pure component spectrum where the concentration of the component is known, it is possible to quantify this component in all the samples included in the PARAFAC model. This is accomplished without having to perform any kind of peak integration, which would be difficult in a case where no selective peak could be identified as originating from the chemical component in question. This is considered a tremendous improvement compared to the traditional way of working with NMR data of complex mixtures.

By the example presented here we hope to illustrate the enormous future possibilities for the *multi-way* analysis of complex 2D-NMR data with applications in the fields of metabonomics and quality control of food (bromatonomics).

Acknowledgments

The authors wish to thank The Danish Dairy Foundation, The Ministry of Food, Agriculture and Fisheries and the Danish Research Councils for generous support to our NMR project.

References

1 Baunsgaard D, Nørgaard L, Godshall MA, Specific screening for color precursors and colorants in beet and cane sugar liquors in relation to model colorants using spectrofluorometry evaluated by hplc and multiway data analysis, *J Agric Food Chem*, 2001, **49**, 1687-1694.

2 Booksh KS, Jiji RD, Andersson GG, Multi-way modeling of non-linear data: Practical application to excitation-emission matrix fluorometry of aqueous samples, *Abstracts of Papers of the American Chemical Society*, 1999, **217**, 065-ANYL

3 Bro R, Multi-way Analysis in the Food Industry. Models, Algorithms, and Applications. Ph.D. thesis, University of Amsterdam (NL), http://www.mli.kvl.dk/staff/foodtech/brothesis.pdf, 1998.

4 Bro R, Kiers HAL, A new efficient method for determining the number of components in PARAFAC models, *J Chemom*, 2003, **17**, 274-286.

5 Dyrby M, Baunsgaard D, Bro R, Engelsen SB, Multiway chemometric analysis of the metabolic response to toxins monitored by NMR, *Chemom Intell Lab Syst*, 2004, in press.

6 Dyrby M, Peteresen M, Whittaker AD, Lambert L, Nørgaard L, Bro R, Engelsen SB, Analysis of lipoproteins using 2D diffusion-edited NMR spectroscopy and multi-way chemometrics, *Anal Chim Acta*, 2004, in press.

7 Eastman HT, Krzanowski WJ, Cross-validatory Choice of the Number of Components From a Principal Component Analysis, *Technometrics*, 1982, **24**, 73-77.

8 Faber NM, Ferre J, Boque R, Kalivas JH, Second-order bilinear calibration: the effects of vectorising the data matrices of the calibration set, *Chemom Intell Lab Syst*, 2002, **63**, 107-116.

9 Gemperline P, Puxty G, Maeder M, Walker D, Tarczynski F, Bosserman M, Calibration-free estimates of batch process yields and detection of process upsets using in situ spectroscopic measurements and nonisothermal kinetic models: 4-(dimethylamino)pyridine-catalyzed esterification of butanol, *Anal Chem*, 2004, **76**, 2575-2582.

10 Gurden SP, Westerhuis JA, Bijlsma S, Smilde AK, Modelling of spectroscopic batch process data using grey models to incorporate external information, *J Chemom*, 2001, **15**, 101-121.

11 Harshman RA, Foundations of the PARAFAC procedure: Models and conditions for an 'explanatory' multi-modal factor analysis, *UCLA working papers in phonetics,* 1970, **16**, 1-84.

12 Harshman RA, De Sarbo WS, An application of PARAFAC to a small sample problem, demonstrating preprocessing, orthogonality constraints, and split-half diagnostic techniques, *Research Methods for Multimode Data Analysis,* (Eds. Law, HG, Snyder, CW, Hattie, JA, and McDonald, RP), Praeger Special Studies, New York, 1984, 602-642.

13 Hotelling H, Analysis of a complex of statistical variables into principal components, *Journal Of Educational Psychology,* 1933, **24**, 417-441.

14 Kruskal JB, Three-way arrays: rank and uniqueness of trilinear decompositions with applications to arithmetic complexity and statistics, *Linear Algebra and its Applications,* 1977, **18**, 95-138.

15 Laster L, Statistical background of methods of principal component analysis, *J Periodontol,* 1967, **38**, Suppl-66.

16 Nikolajsen RPH, Booksh KS, Hansen AM, Bro R, Quantifying catecholamines using multi-way kinetic modelling, *Anal Chim Acta,* 2003, **475**, 137-150.

17 Pearson K, On lines and planes of closest fit to points in space, *Philosophical Magazine,* 1901, **2**, 559-572.

18 Selli E, Zaccaria C, Sena F, Tomasi G, Bidoglio G, Application of multi-way models to the time-resolved fluorescence of polycyclic aromatic hydrocarbons mixtures in water, *Water Research,* 2004, **38**, 2269-2276.

19 Sidiropoulos ND, Bro R, On the Uniqueness of Multilinear Decomposition of N-way Arrays, *J Chemom,* 2000, **14**, 229-239.

20 Smilde AK, Bro R, Geladi P, *Multi-way analysis. Applications in the chemical sciences.* John Wiley & Sons Ltd., UK, 2004.

21 Vega-Montoto L, Wentzell PD, Maximum likelihood parallel factor analysis (MLPARAFAC), *J Chemom,* 2003, **17**, 237-253.

22 Wang YD, Borgen OS, Kowalski BR, Gu M, Turecek F, Advances in 2nd-order calibration, *J Chemom,* 1993, **7**, 117-130.

THE EFFECT OF POROUS STRUCTURE OF RICE ON THE HYDRATION RATE
INVESTIGATED BY MRI

Aleš Mohorič[1,2], John van Duynhoven[3], Gerard van Dalen[3], Frank Vergeldt[1], Edo
Gerkema[1], Adrie de Jager[1], Henk Van As[1]

[1]Laboratory of Biophysics and Wageningen NMR Centre, Wageningen University,
Dreienlaan 3, 6703 HA Wageningen, The Netherlands
[2]Faculty of mathematics and physics, University of Ljubljana, Jadranska 19, 1000
Ljubljana, Slovenia
[3]Unilever Research Vlaardingen, Olivier van Noortlaan, PO Box 114, 3130 AC
Vlaardingen, The Netherlands

1 INTRODUCTION

The rate of cooking is an important parameter in the consumer appreciation of instant rice.
When internal structure is changed (gelatinisation degree or porous structure), the cooking
time decreases. We used NMR imaging to investigate the hydration process on a fine
spatial scale and in real time in types of rice undergone different treatment. The motivation
is to understand the effect of internal structure on hydration process.

When a rice grain is subjected to heat and pressure in the presence of water the starch
in granules will change its structure (gelatinisation). During the subsequent cooling and
drying the molecular structure may change again (retrogradation, formation of V-type lipid
complex) and pores can form. The resulting structure depends on the process. Since the
parameter space is overwhelming, suitable methods are needed to evaluate the outcome[1].
An example of thermal treatment effect on the porous structure of a grain is shown in
Figure 1 where central slices of X-ray tomographs show an extensive porous structure in
the processed type. Native starch is in the form of semi-crystalline granules. During
gelatinisation the granules are melted and mostly amorphous structure with possible
retrograded portion forms[2]. Both molecular and porous structure have an effect on the
cooking rate.

Figure 1 *Central slices of XRT scans of native (left) and processed (right) rice grains.
Raw starch is compact and in form of semi-crystalline granules. Processed
grain exhibits extensive porous structure.*

The cooking rate can be evaluated by the analysis of magnetic resonance images of rice during cooking. Signal of a pixel is proportional to moisture content[3] and thus a suitable detector for the presence of water.

2 EXPERIMENTAL

3D maps of rice during cooking were made in rapid succession using RARE imaging sequence. The process was followed in real time with a 3D map scanned in 1 min approximately every 1.5 min. If quenching is applied to stop the cooking process, longer measuring time is possible, but the short-time effect of quenching is not known. The poor S/N method is used because it is fast and the scanning rate is important in a fast process such as cooking[3]. If longer measuring time is used, water distribution will change significantly during the scan and the resulting image is not an adequate representation of moisture profile. The spatial resolution is important as well since structure anomalies such as cracks or pores have strong influence on the hydration rate and need to be resolved in order to properly evaluate the rate.

Because the RARE method needs to be fast and the relaxation times of proton signal in the grain are short (typically on the order of 10 ms)[4,5], the centre of k-space must be covered with the first echoes of the RARE echo train and the centred-out phase encoding[6] was used. The second phase encoding dimension (denoted here as k_{slice}) was covered linearly from minimum to maximum as shown in Figure 2. Repetition time of 0.5 s was used. This effectively destroys free water signal and enhances the signal of hydrated starch. In the echo train 16 echoes were recorded with the first echo forming 6 ms after the 90° excitation pulse. The resolution of the images was 128×32×16 with a field of view of 15×5×5 mm^3. The number of averages of 4 was enough for an acceptable signal to noise ratio. The images of a single rice grain during cooking were recorded on a 0.7 T Bruker Avance MRI system.

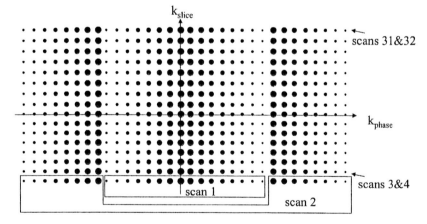

Figure 2 *Centred-out phase encoding used in the study. The size of the circle indicates the order of the echo: the largest circle is the first echo and the smallest is the last echo of the echo-train. The read-out direction of k-space is perpendicular to the plane of the diagram.*

Air was heated outside the probe and was led past the tube containing grain and excess water. Temperature was set to 90 °C, just above the gelatinisation temperature of starch, and was adjusted in a dummy experiment where the temperature was measured on site.

3 RESULTS AND DISCUSSION

The results of MRI scans are shown in Figure 3, where a series of central slices show different types of grain during cooking. The images reveal the internal structure of the grain, as water diffuses into the grain. The structure disappears as the cooking process continues and the starch gelatinises and swells. The moisture front is clearly visible and can be used to evaluate the rate of hydration and the process governing it, namely, Fickian diffusion or water demand[7] driven imbibition. Cracks developed in un-processed grain during storage period in a dry compartment and these are clearly visible on MR images. Hydration along the cracks is significantly faster (around 25 min) than in the bulk starch (around 30 min). Using lower resolution this effect may not be visible. Pores in processed grain can be clearly distinguished even at this relatively low resolution. Some pores are hydrated even before the cooking process is initiated. This happens when pores are connected with the surface and an escape route for the trapped air exists.

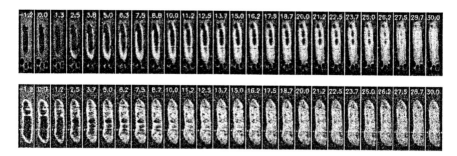

Figure 3   *Cooking of native (top) and porous (bottom) rice. Shown is the central slice of the grain for different times (indicated at the top in minutes) during cooking. In the top series, a few cracks are visible and along them the hydration is faster. In the bottom series pores can be distinguished. They disappear as starch gelatinises and swells.*

To make the evaluation of hydration rate easier, 1D plots of moisture content across the centre of the central slice were collected in a diagram for different times during cooking. From this plots, shown in Figure 4, different models simulating the hydration, derive parameters[8]. We were able to distinguish between several different types of hydration:

- difficult water uptake at the edge, initially Fickian diffusion, later high water demand driven diffusion, pore structure (if present) plays no important role,
- difficult water uptake at the edge, Fickian diffusion during complete interval of water uptake, pore structure (if present) plays no important role,

- very easy water uptake at the edge as well as in the interior of the rice grains, very fast water uptake, pore structure is connected and open to the outside, pore structure facilitates water uptake
- easy water uptake at the edge, Fickian diffusion during complete interval of water uptake, pore structure plays no important role

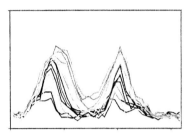 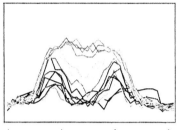

Figure 4   *1D plots across the centre of the central slice showing the moisture profile of the grain during cooking. On the left, the moisture front is sharper which indicates imbibition and on the right, the hydration process is of the Fickian diffusion type.*

The signal integrated over the volume of the grain gives us the hydration rate and also the time of cooking. The information in these plots is equivalent to gravimetric results when the NMR signal is properly normalized[3]. Such curves are shown in Figure 5. The hydration rate is faster in porous structure and the results match with the gravimetric measurements. If a model, describing the increase in moisture content, $m=m_0[1-\exp(-t/T)]$ is used, the characteristic hydration time T of native grain is 21 min and 12 min for the processed grain.

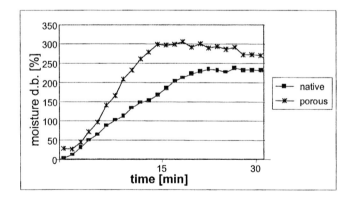

Figure 5   *The signal integrated over the grain and normalized to the moisture content of the gravimetric measurement gives us the increase in moisture content with the time of cooking. The native grain hydrates longer (around 20 min) than the porous type, which is fully hydrated in about half the time.*

The results are expected: thermal treatment of rice, which results in porous, gelatinised dry starch, will produce faster cooking rice. The optimisation of the treatment can be supported by the presented technique. However, the consumer appreciation or the taste of rice is not determined in such a study.

References

1   B. Hills, Magnetic Resonance Imaging in Food Science, 1998, Wiley.
2   M.A. Kaddus Miah, A. Haque, M.P. Douglass, B. Clarke, *Int. J. Food Sci. and Techn.*, 2002, **37**, 527.
3   A. Mohorič, F. Vergeldt, E. Gerkema, A. de Jager, J. van Duynhoven, G. van Dalen, H. Van As, *J. Magn. Reson., In Press.*
4   S. Takeuchi, M. Maeda, Y. Gomi, M. Fukuoka, H. Watanabe, *J. Food Eng.*, 1997, **33**, 281.
5   A.G.F. Stapley, T.M. Hyde, L.F. Gladden, P.J. Fryer, *Int. J. of Food Sci. and Techn.*, 1997, **32**, 355-375.
6   T.W.J. Scheenen, D. van Dusschoten, P.A. de Jager, H. Van As, *J. Magn. Reson.*, 2000, **142**, 207.
7   H. Watanabe, M. Fukuoka, A. Tomiya, T. Mihori, *J. Food Eng.*, 2001, **49**, 1.
8   L.R. van den Doel, A. Mohorič, F. Vergeldt, J. van Duynhoven, H. Blonk, G. van Dalen, H. Van As, L.J. van Vliet, *submitted to J. Comp. Phys.*

OIL CONTENT VARIATIONS IN SUNFLOWER ACCORDING TO THE GRAIN ORIENTATION COMPARATIVELY MEASURED BY SPIN-ECHO AND CWFP SEQUENCES

A.D. Dumón,[1] M.G. Albert,[1,2] and D.J. Pusiol [1,2]

[1]SpinLock SRL, C. de Arenales 1020, X5000GZU Córdoba, Argentina
[2]FaMAF, Universidad Nacional de Córdoba and CONICET, Ciudad Universitaria X5016LAE Córdoba, Argentina

1 INTRODUCTION

Continuous Wave Magnetic Resonance[1] has been used during many years to select seeds of high oil content in a rapid and non-destructive way. Even though this technique is non-destructive and quite rapid, the measurements require the seeds to be previously dried and weighed. Pulse Magnetic Resonance (P-NMR)[2] has been used as an evolutionary technique regarding the previous one since the samples do not need to be dried beforehand. Additionally, magnetic field high homogeneity required by continuous wave spectrometers is no longer a critical parameter in P-NMR spectrometers. In this way, the cost of the equipment is lower.

At the present, three variants of the P-NMR method were developed: i) a simple signal amplitude measurement that follows the application of a single radio frequency pulse[3]; ii) the known solid liquid relation method measured through the spin-echo sequence, S-E[4]; and, iii) the recently published method of Continuous Wave Free Precession, CWFP[5]. In the first method, the contributions of each seed component to the Free Induction Decay (FID) that follows the $\pi/2$ radio frequency pulse[2] are measured at different evolution times of the aforesaid signal. The signal of the solid components decreases quickly, in less than seventy microseconds; at this point the signal corresponds to water and oil contents. Then, the signal is measured again in a time of a millisecond order, at this time the moisture signal has totally relaxed. In this way, the millisecond measurement corresponds to the total amount of oil present in the sample. The moisture signal corresponds to the difference between the FID amplitudes of seventy microseconds and a millisecond respectively. This method requires high homogeneity in the magnetic field since the inhomogeneities in the magnetic field distort FID measurements particularly at times over a hundred microseconds.

The second method does not need very high homogeneity in the magnetic field at the expense of a slight loss in the results security. This method is based on the FID simultaneous measurement of the moisture-oil ratio at seventy microseconds from the end of the RF pulse. The total amount of oil is measured in the spin-echo amplitude, which follows a spin-echo sequence. Once the $\pi/2$ pulse is applied and the FID has occurred at a following τ time, a second pulse of refocusing is applied; this pulse has a double width regarding the first one. After another τ time interval has occurred, the spin-echo appears. The echo amplitude is given by:

$$A(2\tau) = A_0.exp[-(2\tau)/T_2 + D]$$

where A_0 is the FID amplitude immediately after the first pulse has finished, D is a term related to the auto diffusion coefficient and T_2 is the decay time of the transversal signal. Due to the fact that in oilseeds the oil T_2 is about 10 times larger than the moisture T_2, it is possible to pick a τ value in which the oil signal persists when the moisture signal has completely decayed. In this way, after calibration the results of the spin-echo signal will be a measurement of the total amount of oil, while its difference in the FID signal digitized at seventy microseconds will represent the total amount of moisture. It has been proved that, within certain limits, the measurement obtained from spin echo is independent from the magnetic field inhomogeneities.

An important improvement in both methods has been recently published[5]. This improvement consists of applying successive $\pi/2$ pulses to the sample so that the interference between the FID and the spin echo signals becomes constructive. Just like the previous methods, the total moisture and oil signals are taken at seventy microseconds at the end of the first $\pi/2$ pulse. After the time needed to reach the stationary state imposed by the Continuous Wave Free Precession (CWFP) condition, many signals made up of FID and spin echo can be obtained. As a result, these signals' average can be rapidly calculated to improve the signal-noise ratio of at least one magnitude order regarding the previous methods.

The simultaneous oil, moisture and solid matter determination is based on the fact that the protons T_2 spin-spin relaxation time is much longer in oil than in protein, carbohydrates and water. This fact allows separated measurements of the four main components of oilseeds. The transverse relaxation signal of each component is digitalized and turned into the oil individual percentage. To achieve these results, the sample is previously weighed and a linear calibration curve is used. In order to determine the oil content in an accurate way, it is necessary to find the appropriate experimental conditions so that the oil percentage does not depend on the spin-lattice relaxation time (T_1), the spin-spin relaxation time (T_2), the angular position of the seeds in the sample tube, the thickness of the tube walls and the sample volume. It is also of great importance that the signal assigned to the oil does not present solids (proteins and carbohydrates) and/or moisture components.

In commercial spectrometers, only about 10 gr. of the total amount of sunflower in the sample can be analyzed in a determination. Moreover, a commercial unit that is generally accepted corresponds to the loading capacity of a transport truck. In practice, the inhomogeneities of a commercial unit imply that a single sample of 10 gr. is not necessarily represented in an accurate way. On the other hand, the aforesaid NMR measurement methods are also sensitive to the seeds orientation relative the magnetic field, therefore the sample amount is what assures a sample of higher homogeneity and representation. According to the current NMR technology, the oil content measurement repeatability in a single sample is of the order of 0.5%. In practice, the sample measurements of a single commercial unit vary from 1 to 2%. The purpose of this work is to study the influence of moisture-fat matter relation of sunflower seeds in the measurements according to i) its orientation in the magnetic field and ii) the inhomogeneities in a volumetric sample. This study also compares the relative performances of S-E and CWFP techniques. It is intended to achieve measurement results that represent a commercial unit and a precision compatible with the inherent repeatability of the current NMR technique development. In order to achieve this, each analysis time needs to be diminished in order to increase the sample amount and therefore reach a more accurate statistic.

2 EXPERIMENTAL

The measurements were performed in the SpinLock SE-100 spectrometer which works at 15 MHz Larmor frequency. The magnetic field homogeneity in the sample volume is estimated to be 20 ppm. The sample holder tubes were made of Teflon, of 35 mm in diameter and 40 mm in length. They have a capacity of 11 gr. of sunflower seeds to keep a lineal measurement in a range of 0.2%. The seeds were supplied by General Deheza Oil Plant (AGD SA). They form 15 lots of material taken from the normal production in the province of Cordoba. The sample temperature was not controlled. The measurements were taken from the spectrometer magnet at room temperature without heating or any other type of temperature control. Oil seeds content of each lot were previously measured in the AGD Laboratory in a Bruker Minispec spectrometer which works at 10 MHz Larmor frequency.

Figure 1 shows a graphic that represents both spectrometers' linearity; the S-E method was used for both spectrometers. An excellent agreement between them in the order of 1.1% is noticed.

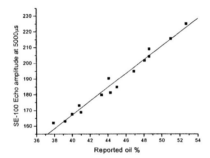

Figure 1   *Correlation between measurements performed by a Bruker Minispec spectrometer (axis x), with those made by a SpinLock SE-100 spectrometer (axis y)*

Likewise, Table 2 shows the reproducibility of the SR15 Magnetic spectrometer results, working in the same sample, using both measurement sequences: CWFP in column 2 vs. S-E in column 3. These sequences show a sample distribution of 0.063 and 0.344 respectively. Both samples mid values are 37.7 and 38.1. From this preliminary study we conclude that the same equipment gives results of higher repeatability when using CWFP instead of S-E sequence, although both methods share the same mid values.

Table 1   *Statistical dispersion of oil content in a sunflower sample, measured by the same spectrometer (SpinLock SE-100), 10 times in each sequence*

Measurement N°	CWFP	S-E
1	37.8	37.7
2	37.7	38.3
3	37.8	37.9
4	37.8	38.0
5	37.8	38.6
6	37.7	38.0
7	37.7	38.5
8	37.8	38.3
9	37.8	37.5
10	37.9	38.1
Average	*37.8*	*38.1*
Std. Dev.	*0.063*	*0.344*

Measurements were taken from different orientations of the sunflower seed main axis relative the magnetic field in order to verify the origin of the signal magnitude dispersions. For that purpose, three sample holder tubes were filled with seeds of the same lot and they were placed according to three orientations in relation to the axis of the tube: i) one in which the seeds major axis is perpendicular to the axis of the tube –from now on: "in horizontal position", ii) another which follows the direction of the tube axis or "placed in vertical position", and iii) the other in which the seeds are randomly distributed relative their relative orientations. The sample holder tube axis is perpendicularly placed to the magnetic field which, in our spectrometer, has a horizontal direction.

The measurements of the oil signal amplitude by the CWFP sequence allow us to observe the following:

1. On average the measured NMR signal is larger when the seeds are oriented according to the major axis in the magnetic field direction than when they are oriented maintaining that axis perpendicular to the magnetic field. Moreover, when the seeds relative orientations are random, a signal amplitude of mid magnitude is measured.

2. We notice that by placing the seeds in horizontal position and rotating the sample holder tube in a perpendicular direction to the magnetic field, a complete rotation of oil signal amplitude oscillates in a sinusoidal form, see Figure 2. When the seeds oriented in horizontal position are aligned longitudinally to the external field, the maximum signal is obtained and when they are aligned transversally, the minimum signal is obtained.

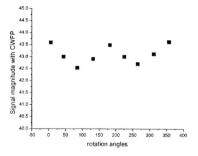

Figure 2   *Signal magnitude variations by CWFP from a complete rotation of the sample holder tube. The total variation between the extremes is 1.2%.*

The measurements were repeated in three samples of the same lot, at this time the rotation angle was varied from 0° to 180° according to the perpendicular direction. It is noticed the same functional behavior of the oil amplitude in the rotations and some magnitude differences in each seed signal. Such differences are also noticed when the seeds are oriented in a vertical position.

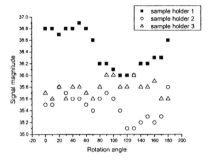

Figure 3   *Variation of signal intensity by CWFP during a horizontal seeds rotation of 180°. The dispersion between the maximum and minimum values is 2%..*

The same seeds were placed in vertical position using the same sequence. In this case the influence of the position is not that obvious, this can be seen in Figure 4.

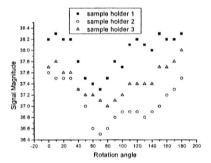

Figure 4   *Amplitude variation in vertical seeds of 180° rotation by CWFP*

In Figures 3 and 4, it is observed that the difference in fat matter content values of each seed does not vary. In both cases, the seeds of the tube 1 have a larger signal than those of the tubes 3 and 2 respectively. Notice also that, in the three cases, larger signal amplitude is observed in seeds that are placed horizontally relative those placed vertically. This difference reaches up to 1% of the oil percentage.

In practice, the seeds in the sample holder tube are expected to be placed randomly oriented in relation to the magnetic field. This implies that the measured signals magnitudes are determined by the average of all the relative orientations.

Figure 5 shows the behavior of oil contents obtained from the three aforesaid samples which present random seeds distributions in different angles of the sample holder tube rotation. It is noticed that the relation in Figures 3 and 4 is no longer the same with these new results. In particular, the amplitudes measured in the sample holder tubes 1 and 3 do not have the same intensity relation than the one observed in Figures 3 and 4, even when the orientation dispersion is diminished to the order of 1%. However, the angular dispersion obtained from the sample of the sample holder tube 2 presents a larger angular variation. This effect could be achieved due to a smaller disorganization in the seeds orientations produced during the test preparation.

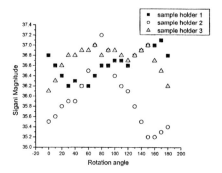

Figure 5   *Signal magnitude variations by CWFP in seeds that are randomly oriented. It corresponds to a 180° rotation*

Figure 6 shows the results obtained from the same seeds, introducing a second mechanical disorganization into each sample holder tube. Although the mid values of the total percentage of each sample oil compared to both mechanical treatments remain within the +/- 0.3 % error, it is observed that the angular dependence of each sample can change in a +/- 0.5 % range.

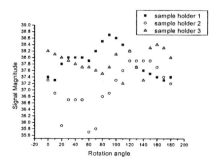

Figure 6   *Signal magnitude variation in a 180° rotation after a second random orientation using CWFP*

In order to compare the influence of the seeds orientations relative the magnetic field according to CWFP and Spin-Echo techniques, the same measurements in the same conditions were carried out, using in this case the Spin-Echo technique. Figs. 7, 8 and 9 show the signal magnitude differences in horizontal, vertical and random orientations respectively.

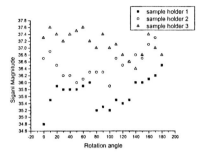

Figure 7 *Signal amplitude variations in seeds horizontally oriented during a 180° rotation using S-E*

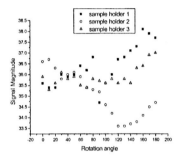

Figure 8 *Signal magnitude variations in seeds vertically oriented during a 180° rotation using S-E*

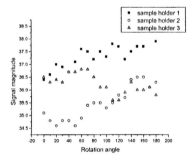

Figure 9 *Signal magnitude variations in seeds randomly oriented during a 180° rotation using S-E*

After comparing the results obtained from both measure sequences for each orientation, it is noticed that the CWFP sequence shows wider variations in the oil signal magnitude than those obtained from the Spin-Echo sequence.

3 CONCLUSIONS

The results of sunflower oil percentage measurements by CWFP sequence presents the following characteristics regarding measurements by Spin-Echo sequence:
1) Greater accuracy.
2) Greater angular dependence on seeds orientation relative the magnetic field.

The explanation of the orientation dependence of oil percentage obtained from both techniques is based on the comparative analysis of the signals measured in each technique. In fact, the oil signal generated by a Spin-Echo sequence is mainly determined by the spin echo amplitude while the signal measured by CWFP technique is a composition of the FID signal and the Spin-Echo signal, formed by a τ of at least one order of magnitude smaller than the one shown by the S-E sequence. On the other hand, like water molecules, the seeds oil is made up of molecules that present two main types of mobility: one of liquid nature which is concentrated in volumes far from solid fibers, proteins and macromolecules, and the other which is bonded to molecules of solid nature. The former has high mobility and therefore its contribution to the oil signal is highly independent from its magnetic field orientations since this thermal agitation is enough to average the local fields. The latter, which has restricted mobility due to macromolecules with very little orientation diffusion, presents important contributions to the CWFP signal for both the FID and the echo. Thus, by picking a τ time between the exciting pulse and the proper refocusing pulse (it is usually τ ~ 5 ms in the sunflower), the spin echo of the SE sequence contains mainly NMR signal from molecules in liquid state since these have longer T_2. On the other hand, a greater contribution of low mobility oil molecules to the CWFP signal is expected since the separations between pulses are taken at a hundred of microsecond order when the moisture signal has completely relaxed but yet the oil molecules signal bond to the fibers has not. It is possible that it is not a characteristic effect of other oilseeds such as the soy in which the fiber molecules are not oriented towards preferential directions.

On the other hand, the results obtained from CWFP sequence are more repetitive and accurate. This characteristic is achieved when a shorter τ time is defined. As a result, a greater amount of signals are possible to digitalize and therefore their average is more efficient than the average obtained from the SL sequence.

Moreover, it is evident that in both cases the amount of seeds of 10 gr. in a sample is not enough to obtain an orientation accurate statistic.

The development of the future work considers the integration of both sequences in order to study their advantages.

References

1 A. Abragam, *The Principles of Nuclear Magnetism,* Clarendon Press, Oxford, 1996.
2 C.P. Slichter, *Principles of Magnetic Resonance*, Spinger-Verlag, Heidelberg, 1990.
3 P.N. Tiwari, P.N. Gambhir y T. S. Rajan, *Rapid and nondestructive determination of seed oil by pulsed nuclear magnetic resonance technique,* J. Am. Oil Chem. Soc., 1974, **51**, 104.
4 P.N. Tiwari y W. Buró, J. Amer. Oil Chem. Soc., 1980, **57**, 119.
5 R.B.V.Azeredo, L.A.Colnago, A.A.Souza, M.Engelsberg, *"Continuous wave free precession: Practical analytical tool for low-resolution nuclear magnetic resonance measurements",* Analytica Chímica Acta, 2003, **478**, 313.

COMPREHENSIVE PHASE COMPOSITIONAL ANALYSIS OF LIPIDS IN FOODS
BY MEANS OF TIME DOMAIN NMR

E. Trezza, A. Haiduc, E.C. Roijers, G.J.W. Goudappel, J.P.M. van Duynhoven

Unilever R&D Vlaardingen. P.O. Box 114, 3130AC Vlaardingen, The Netherlands.

1 INTRODUCTION

1.1 Lipid Phase Composition in Foods

In food science and technology, assessment of phase compositional behavior of lipids is critical for understanding many product properties and for effective process control. Well known examples of such properties are the texture (hardness) and melting behaviour of fat-continuous products as margarines and chocolate.[1] Also in many water-continuous products, the phase-behaviour of lipid components is known to have a major impact of food properties.[2]
Critical lipid phase-compositional parameters for the food technologist are crystal polymorphism and solid (fat) content (SFC). The presence of mesoscopic (semi-solid) lipid form has also been recognised as a parameter of interest. The assessment of crystal polymorphism in food lipids has traditionally been the domain of X-ray diffraction and DSC. The Solid Fat Content in fat blends was traditionally measured by dilatometry,[3] until almost three decades ago benchtop NMR technology was introduced for this purpose by Unilever and Bruker.[4] The assessment of semi-solid phases in blends and products has typically been the domain of NMR.[5]
The current SFC NMR methods still use rather crude acquisition and data treatments of the Free Induction Decay (FID). However, since the introduction of this NMR method in the 70s, the electronic and computational specifications of commercial benchtop NMR equipment have improved dramatically and current SFC methods do not fully exploit this. Several recent NMR studies have shown that benchtop NMR equipment can be used to obtain detailed phase-compositional information on complex systems.
In the detergent area, several applications have been described that show that the phase behaviour of surfactants can be assessed in a detailed manner by curve fitting of the time domain data generated by benchtop NMR spectrometers.[6] It has also been demonstrated [7] that Curve Fitting of rapidly sampled transverse NMR relaxation decay curves enables the quantitative determination of the mesomorphic state of polysaccharides. Also examples are known where NMR relaxometry was used to assess semi-solid behaviour of lipids.
Recently, we established the feasibility of extracting a range of phase compositional parameters of lipids from NMR relaxation decays.[8] This encouraged us to embark on a

systematic study on Transverse Relaxation Decay Deconvolution (TRDD). First, we assessed the variety of relaxation decay lineshapes for different triglyceride polymorphs. The lineshapes were used to compose a multi-component function that can be used to describe systems with contributions from liquid, semi-solid and solid phase, where the latter can exist in different crystal polymorphs. Accuracy and precision are critical for acceptance of methods within the industrial practice, hence extensive validation exercises were carried out. Method accuracy also often referred to as 'method trueness' is here assessed as the (statistical) equivalence with the classical 'Direct SFC' method. Method precision often referred to as 'standard deviation under practical circumstances' is determined via a validation exercise. Finally, in order to demonstrate the suitability of the NMR method for practical purposes, we will present several real-life industrial applications.

2 MATERIALS AND METHODS

2.1 Materials

    2.1.1 Pure Triglycerides. Lineshapes were defined using eleven pure (>99%) triglycerides that were known to be exist for relatively long times in either the ∀, γ, Ⅎ, or Ⅎ' polymorphs.
    2.1.2 Commercial Fat Blends. Fat blends that were extracted from 40 commercial margarine products. These products had a wide variation in fat blend composition and N-lines.
    2.1.3 Oil-in-Water Emulsions. Emulsion premixes were prepared at 50 °C and homogenised (app. 20 min). Homogenisation was performed using a lab-scale homogenise, and is preceded by turraxing the mixture for 2 min. Emulsions were filled in 100 mL tubs, sealed, and stored at 5 °C until further analysis. The samples were cooled in the fridge in 1-2 h.

2.2 Data Acquisition

All NMR data were measured on commercial Bruker MQ20 NMR spectrometers, operating at 20 MHz for ^{1}H. The instruments were equipped with 10 mm variable temperature probeheads (PH20 10-25/R/V). A home-written pulse sequence was used, combining a Free Induction Decay (FID) and a Carr-Purcell-Meiboom-Gill (CPMG) decay. The FID was sampled for 100 µs, with a frequency of 2.5 MHz. The 180^0 pulses of the CPMG sequence were spaced by 200 µs. Both the FID and CPMG signals were acquired in phase-sensitive mode. The 90^0 pulse was typically in the range of 1 to 2.5 µs. All samples were measured at 20 °C, with a recycle delay of 10 s, unless mentioned otherwise. In order to avoid artefacts in the CPMG signal (i.e. wiggles on the first data points) care was exerted to minimise B_0 and B_1 inhomogeneity effects. This involved the use of small sample heights (app. 1 cm), and a well-adjusted B_0 homogeneity. Typically, the B_0 homogeneity was such that the t $_{1/2}$ of an oil sample was >1 ms.

2.3 Data Analysis

The complete data-analysis procedure involved several pre-processing steps: subtraction of background signal, correction for field-inhomogeneity and phasing. Subsequently, the

NMR data were fitted with pre-defined functions. All procedures were programmed in Matlab®. The FIDCPMG decay was fitted with functions that correspond to Gaussian, Pake [9] and exponential lineshapes:

$$FIDCPMG \ (t) = \sum_i^{l1} L_1(i) * e^{t/T_{2,i}} + \sum_j^{l2} L_2(j) * e^{t/T_{2,j}} + \sum_k^{g} G(k) * e^{-(t/T_{2,k})^2} +$$

$$\sum_l^{p} P(l) * \exp(\frac{66.5^2 * t^2}{2}) * \frac{\sin(Bt)}{Bt} + \sum_n^{l3} L_3(n) * e^{-t/T_{2,n}}$$

The fitting procedures were performed by means of a non-negative least-squares approach.[10]

3   RESULTS

3.1   Definition of Lineshapes for Common Lipid Polymorphs

In order to get a good quantitative description of transverse relaxation decays of lipid crystal polymorphs we have measured a range of representative examples. In order to avoid the dead-time which is normally present in FID's, we acquired these lineshapes by means of a solid echo sequence.[11] In total 4 lineshapes for α polymorphs were recorded, 6 for the β and 3 for the β' form. For the rather uncommon γ polymorph, two decays were recorded (SOS and POP). The collection of lineshapes obtained for the different polymorphs were fitted with the previously described functions using constrained parameters. We observed that the variation *between* lineshapes of β and β' was in the same order of magnitude as the variation *within* the lineshapes of the respective polymorphs. Hence, we concluded that these polymorphs cannot be discriminated in a reliable manner by transverse relaxation decays and a single set of parameters has been used for both species.

3.2   Multi-component Fitting of Experimental Lineshapes

The lineshapes studies presented so far contain the contribution of one crystal polymorph only. In practice, we encounter heterogeneous systems such as blends or semi-product formulations. Such systems can contain the aforementioned lipid polymorphs, but also liquid oil and/or water, and states with intermediate mobility, here referred to as 'semi-solid'. We would like to assess the contributions of these different phases from experimental relaxation decays. In our fitting procedure we fit the contributions of the polymorphs β and β' by one single lineshape, denoted as $\beta^{()}$. It was found that in heterogeneous systems the pure lineshapes of crystal polymorphs were not adequate in describing experimental decays. We found that adding a rapidly decaying exponential function ($T_2 < 45$ µs) was sufficient of obtain good fits. We attribute this contribution to increased molecular mobility in the solid lipid phase due to crystal imperfections and inclusion of impurities.

The abundance of parameters in this complex multi-component function requires that these are fitted within defined constraints. The constraints regarding the actual lineshapes are fixed according to their variation between different lipids species. The relative fractions of the different lipid phases are constrained to be non-negative. The overall lineshape was fitted with a Non-Negative-Least-Squares (NNLS) fitting algorithm, which is well suited for non-parametric fits of complex functions.

3.3   Practical Applications of Lineshape Analysis Approach

3.3.1   Crystallisation Kinetic of a Lipid Blend Monitored by XRD and NMR TRDD. The polymorphs formation during crystallisation of a fat blend was measured by NMR TRDD (Figure 1A). The blend was quenched at 5 °C and kept at this temperature during the 2 hours observation time. The TRDD results show that the first crystals formed are in the thermodynamically unstable α form. Together with the α formation, a very fast decaying signal ($T_2 < 0.45$ μs) can be detected (Figure 1A, open symbols) probably marking the beginning of the $\beta^{(\prime)}$ crystals formation. Later, $\beta^{(\prime)}$ crystals can be detected. After one hour all the α form is converted into the more stable $\beta^{(\prime)}$ polymorph. These data are compared with the results of XRD measurement performed on the blend in the same condition (Figure 1B). The XRD analysis confirms a slow conversion from α to β form. According to XRD the crystallisation, process is faster: already 40 % of the sample crystallise within the first measurement time. The α form is only detected for the first thirty minutes. The different kinetic behaviour observed with the NMR method can be explained as this technique investigates the crystallisation behaviour more close to that in the bulk and is more sensitive to detect small crystals.

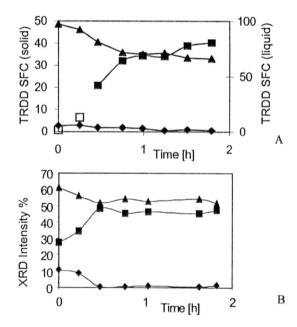

Figure 1 *Phase-behaviour kinetics of blend analysed by NMR (A) and XRD (B). Within the solid fraction α and β polymorphs can be distinguished. Closed symbols indicate unambiguous phase assignment. Open symbols indicate unidentified or ambiguous assignment (see text). The insets refer to XRD data.*

3.3.2 *Lipid Crystallisation Kinetics in Oil-in-Water Emulsions.* So far we presented results of TRDD applied to pure lipids and blends. In the research and product development practice, one would also like to assess phase-composition of lipid-based (semi-)formulated products such as oil-in-water (O/W) emulsions.

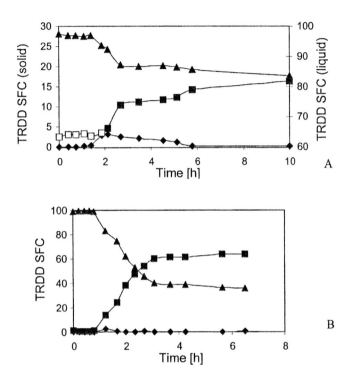

Figure 2 *(A) Crystallisation kinetic of lipid components in an O/W emulsion (B) Crystallisation kinetics of the same lipid blend that was used to prepare the emulsion in (A). Cooling procedure is identical to (A). Triangles: liquid, diamonds: α; squares: β.*

Figure 2A shows the results of an experiment where lipid crystallisation was monitored of an oil-in-water emulsion that was produced at 60 °C and subsequently cooled to 5 °C (0.083 °C /min). The amount of lipid in this formulation was 20%, and to resolve different polymorphs in the relative small contribution of lipid solids is challenging. TRDD was still able to detect the initial formation of the α polymorph, and its conversion into the β form. This finding was in agreement with crystallisation kinetics as monitored by XRD. For comparison, Figure 2B also presents the crystallisation kinetics of the lipid blend that was used for preparation of the emulsion. It is clear that the phase-compositional dynamics can be established with more sensitivity. However, in the blend one observes only a negligible amount of α polymorph, and most lipid seems to crystallise directly in the β form. This demonstrates that crystallisation behaviour of lipids in blends cannot be used to predict kinetics in emulsions. Direct measurement of lipid crystallisation in semi-formulated products usually requires sensitive XRD detectors, but our measurements show that benchtop NMR provides a good alternative.

3.5 Effect of Habit Modifier on Crystallisation Behaviour of an Oil-in-Water Emulsion

Crystal habit modifiers provide an elegant tool to the food technologist to direct lipid crystallisation into a desired polymorph direction. Such compounds typically consist of polymers or surfactant and are normally dosed in small amounts. The efficacy of CHM's is typically assessed by XRD or DSC, but when the lipid blends are formulated in semi-products such as emulsions these techniques are often inadequate. In many cases one then has to take recourse to functional tests like hardness measurements. Here we applied the TRDD method to an oil-in-water emulsion, consisting of 12% dispersed lipid and a continuous phase which was stabilised by proteins. In such a system, the DSC thermogram is blurred by events in the continuous phase. With a state-of-the-art instrument, XRD is still able to measure lipid polymorph transitions, but such an instrument is generally not at the direct disposal of most food technologists. In Figure 3, the polymorph composition of the lipid component in the emulsion is presented, as determined by TRDD. One can observe that with increasing CHM concentration, the amount of crystallised lipid increases, as well as the relative amount of α polymorph. These species were detected at relative small levels and illustrate the capability of TRDD to assess polymorphism in semi-products.

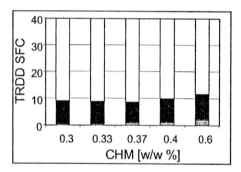

Figure 3   *Effect of crystal habit modifier (CHM) on the phase composition of a protein stabilised oil-in-water emulsion. The bars are arranged in order of increasing amount of CHM (0.3-0.6 w/w %). White, black and grey bars correspond to liquid, β and α polymorphs.*

3.5 Effect of Processing on Phase-Composition in a Semi-Product

Monitoring of lipid crystallisation and polymorphism during processing is a considerable challenge in food technology. Generally, DSC and XRD instrumentation are not suitable to be used in manufacturing environments. Furthermore, the techniques often also fall short when they have to be used for assessment of lipid crystallisation in semi-products. Either the matrix provides too many background signals and/or the lipid fraction is not abundant enough to be detected by these techniques. Benchtop NMR instruments, however, are quite common in manufacturing environments, where they are already routinely used for SFC determinations with one of the 'classical methods'. In Figure 4 we present an example where we assessed lipid crystallisation in a protein-stabilised oil-in-water emulsion on a benchtop instrument with TRDD. As expected, the level of crystalline lipid decreases with temperature, but one can also observe an α to β polymorph conversion. At the highest temperature, one observes melting of lipid crystals and the presence of a semi-solid, which could be a mesomorphic lipid phase. These measurements were confirmed by means of XRD, using a high-end instrument equipped with a non-standard sensitive detector.

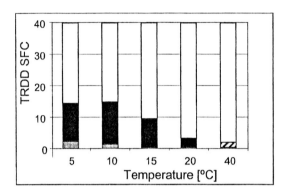

Figure 4   *Phase-behaviour of a protein stabilised oil-in-water emulsion as a function of processing temperature (5, 10, 15, 20, and 40 °C). White, black and grey bars correspond to Liquid, β and α polymorph. Dashed bars correspond to semi-solid.*

4 CONCLUSION

We have presented an NMR approach that is able to assess the phase-composition of lipid-based materials and products in a quantitative manner. When the peculiarities of the classical SFC methods are taken into account, the TRDD method produces equivalent results. The TRDD method is also equally precise as the Direct SFC method, which is generally appreciated for its high precision in the field. The Direct SFC method requires the calibration of an F-factor, whereas the TRDD is calibration-free. In addition, the TRDD can also assess lipid polymorphism.

Several examples have been presented of the use of the TRDD method to monitor phase-compositional dynamics in real-life problems. Even in (semi-)formulated products such as oil-in-water emulsions, TRDD can be used to monitor lipid crystallisation kinetics. Also the subtle effects of crystal habit modifiers and processing parameters can be detected. Traditionally, such measurements require the use of high-end XRD equipment, which are not common in most food laboratories. The results presented here demonstrate that cheap and easy-to-handle benchtop NMR spectrometers cannot only be used to obtain SFC values but also to assess lipid polymorphism. This may render TRDD a powerful tool for research and development laboratories, and it may also have potential for use in manufacturing environment. There, ultrasound is now also routinely deployed for measurement of SFC. However, ultrasound is not able to give the same detailed phase-compositional information, and may even be compromised by crystal polymorphism.

References

1   A. Bot, E. Floter, J.G. Lammers, E. Pelan, "Controlling the texture of spreads", in: *Texture in Food*, Eds. B.M. McKenna, Woodhead, Cambridge, UK, pp. 350-372.
2   D.W. De Bruijne, A. Bot, *In Food Texture: measurement and perception*, Rosenthal, AJ (ED.), Aspen, Gaithersburg, USA, 1999, pp. 185-226.
3   The Solid Fat Index, AOCS Official Method Cd 10-57.
4   K. van Putte, J. van den Enden J.Am.Oil Chem. Soc., 1973, 51, 316.
5   D.J. Le Botlan, L. Ouguerram, *Anal. Chim. Acta* 1997, **349**, 339.
6   S. Nadakatti, *J. Surf. Det.* 1999, **2**(4), 521.
7   I.J. Dries, D. Dusschoten, M.A. Hemminga, E. Linden, *J. Phys. Chem. B* 2000, **104**, 10126.
8   J.P.M. van Duynhoven, I. Dubourg, G.J.W. Goudappel, E. Roijers, *J.A.O.C.S.* 2002, **79**(4) 383.
9   A. Abragam, *The Principles of Nuclear Magnetism*, Oxford, London, 1961.
10 C.L. Lawson, R.J. Hanson, *Solving least squares problems*, N.J. Engewood Cliffs, Prentice Hall, 1974.
11 K. Schmidt-Rohr, H.W. Spiess, *Multidimensional Solid State NMR and Polymers*, Academic Press, 1994.

TWO-DIMENSIONAL LAPLACE INVERSION NMR TECHNIQUE APPLIED TO THE MOLECULAR PROPERTIES OF WATER IN DRY-SALTED MOZZARELLA-TYPE CHEESES WITH VARIOUS SALT CONCENTRATIONS

P.L. Hubbard,[1] P.J. Watkinson,[2] L.K. Creamer,[2] A. Gottwald[1] and P.T. Callaghan[1]

[1]MacDiarmid Institute for Advanced Materials and Nanotechnology, School of Chemical and Physical Sciences, Victoria University of Wellington, Wellington, New Zealand
[2]Fonterra Research Centre, Palmerston North, New Zealand

1 INTRODUCTION

Cheese is an important source of nutrition and, in Western society, is considered to have an appealing flavour and texture. These textural characteristics are affected by the protein:water ratio, pH and extent of proteolysis. Traditional Mozzarella manufacture incorporates a curd-stretching procedure that results in a structure that is akin to that of cooked chicken breast. It has been postulated[1-3] that the casein proteins aggregate and form into strings during the stretching and that the regions surrounding each of the strings contain an aqueous phase containing some of the milk fat globules. This picture was based on electron microscopic examination. Kuo et al.[4] used nuclear magnetic resonance (NMR) relaxation (T_2) to show that the water in Mozzarella cheese could be considered to be in either of two states, or environments; one resembled free water, whereas the other was probably loosely bound to the protein. It was also considered that exchange between the two states was relatively slow.

In the present study, the potentially more powerful relaxation–diffusion correlation technique was tested using four batches of pasta filata Mozzarella cheese, which were made to explore the effect of the quantity of added salt on the characteristics of the water in the cheese as a function of maturation time and test temperature. This correlation technique, in which two-dimensional Laplace inversion NMR is used to provide a map in T_2–D space, is used to analyse the complex multi-exponential behaviour of the relaxation and diffusion rates in heterogeneous systems. It enables us to obtain a plot that is easy to interpret and separates components of a system via their dynamics, revealing additional information by correlating these molecular motions when compared with a one-dimensional technique. Comparison of these results with each other and those of other dairy products has allowed the features of the correlation maps to be assigned. Callaghan et al. described the technique and presented computational details in their 2003 paper[5] and in the previous conference proceedings.[6]

2 MATERIALS AND METHODS

2.1 Samples

The aim of this study was to investigate the properties of both water and fat in cheese as a function of age, temperature and salt concentration using a correlation of characteristic NMR time scales. Four vats of pasta filata Mozzarella-type cheese (moisture 47–49%, fat 20–22%, pH 5.4–5.46 on day 1) were manufactured specifically for this project at a range of added salt concentrations. Vat 1 contained almost no salt (0.09%), Vat 2 contained approximately half the standard quantity (0.88%, low-salt), Vat 3 was a control (1.39%, standard-salt) and Vat 4 contained approximately double the standard quantity (2.11%, high-salt). Nisin, a naturally occurring bacteriocide peptide, was added to all samples to reduce the risk of mould growing in the lower salt samples. The nisin target concentration was 250 IU/g.[7] The large cheese blocks were cut into a number of smaller blocks, individually vacuum packed and stored at 5°C.

A sample was taken from the centre of each cheese block and plugged into an open-ended 10 mm (outer diameter) NMR tube. Tight polytetrafluoroethylene plugs were positioned either side of the sample to minimise moisture loss. Mass measurements taken before and after the experiments showed water evaporation to be negligible. The experiments were repeated on days 3, 7, 12, 29, 42, 63, 97 and 114 for the no-salt and high-salt samples and the following day for the low-salt and standard-salt samples. The samples were placed in the sample tube such that the orientation of the protein fibres ran parallel to the tube in order to standardise the measurement.

2.2 NMR Technique

The NMR experiments were performed on an AVANCE 300 MHz Bruker spectrometer. The two-dimensional correlation maps of spin–spin relaxation time (T_2) versus self-diffusion coefficient (D) were found to give the most insight into the water distribution. The data were collected using a Carr-Purcell-Meiboom-Gill (CPMG) echo train, followed by a Pulsed Gradient STimulated Echo (PGSTE). See Figure 1.

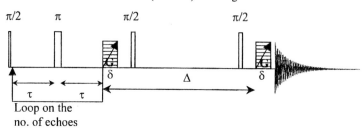

Figure 1   *Pulse sequence for two-dimensional T_2 relaxation–diffusion correlation*

A three-dimensional data file was constructed by collecting data from these pulse sequences in succession. T_2 relaxation data were obtained by looping the CPMG stimulated echo sequence. This was followed by a PGSTE sequence, in which gradients are applied along some chosen axis and ramped. Diffusion was encoded in the z direction (parallel to the tube). The third dimension was the spectral dimension.

Eight scans were collected for each map to increase the signal-to-noise ratio. There were 1024 data points in the spectral dimension and 32 data points in both the relaxation

(T_2) and diffusion dimensions. The time file for the CPMG echo train contained 32 logarithmically spaced values ranging from 1 (0.004 s) to 350 (0.7 s) π pulses. The maximum gradient applied was 8 G/mm and the pulse duration (δ) was 7 ms (7.5 ms including the gradient ramp time). The diffusion time (Δ) was 25 ms and there was no indication of boundary effects causing restricted diffusion.[8]

The two-dimensional correlation maps for each of the cheese samples were collected over time to observe the age dependence of the water and oil fractions. After some preliminary studies at a range of temperatures, the 40°C data were found to reveal more signals from water. Therefore, the majority of the data were collected at 40°C and some were collected at −10°C to assist signal assignment. Different samples were used for each data collection. The samples were left to equilibrate for 30 min before data collection began to allow an equilibrium distribution of oil. Each experiment was 2.5 h in length.

A three-dimensional data file containing FID signals was produced and was Fourier transformed in the spectral dimension. The peak of interest (water or oil) was then integrated to obtain a two-dimensional matrix, which underwent the Laplace inversion to reveal the two-dimensional correlation map.

2.3   Data Processing

The experiments that encode for diffusion and spin relaxation result in exponential decays where the decay rates reveal the parameters of interest. In the case where signals decay exponentially, the inverse Laplace transform is the natural transformation. Ideally, it leads to a sum of well-resolved delta functions, i.e. a diffusion or relaxation spectrum. However, the Laplace transformation is numerically ill defined and unstable to small variations in the input data. As such, special care must be taken when performing the transformation in two dimensions.

The acquisition signals for the two-dimensional experiment correlating T_2 with D are given by:

$$s(t_{i1}, q_{i2}) = \sum \exp(-t_{i1}/T_{2k1}) \cdot \exp(-q_{i2}{}^2 \Delta D_{k2}) X(T_{2k1}, D_{k2}) + E(t_{i1}, q_{i2}) \qquad (1)$$

In the matrix form, both signals can be written as:

$$Y = K_1 X K_2' + E \qquad (2)$$

where K_1 and K'_2 are the known matrices $\exp(-t_{i1}/T_{2k1})$ and $\exp(-q_{i2}{}^2 \Delta D_{k2})$ respectively. E is the error term and represents the noise. Matrix Y can be transformed into a vector and therefore processed using a one-dimensional Laplace inversion.[9] An efficient way to perform such a transform and thus reduce the size of the matrices was first suggested by Song et al.[10]; it uses an algorithm based on Singular Value Decomposition. The distribution X can then be determined by consecutively ordering the matrix rows or columns and performing a non-negative least squares fit.[11] An additional smoothing term weighted by α is required.

$$\chi^2 = \|KX + E\|^2 + \alpha^2 \|X\|^2 \qquad (3)$$

With χ^2 minimised, the two-dimensional Laplace inversion software produces a two-dimensional correlation plot revealing the distribution of relaxation times and diffusion rates.

3   RESULTS

The results in this paper show the two-dimensional correlation maps at 40°C for Mozzarella-type cheeses with various salt concentrations. The sections below detail our findings and give reasons for the current assignments of the two-dimensional Laplace inversion correlation features. The correlation maps show a range from –4 (0.1 ms) to 1 (10 s) for the T_2 relaxation time (the abscissa) and from –4 (1×10^{-13} m^2 s^{-1}) to 1 (1×10^{-8} m^2 s^{-1}) for the diffusion coefficient (the ordinate).

The young cheese samples revealed some intriguing results. Four signals of appreciable intensity were observed in the two-dimensional water maps. Table 1 shows the results for a no-salt Mozzarella-type cheese on day 3, numbered anti-clockwise from the feature with the highest diffusion coefficient. Figure 2 shows the 40°C water data for the four samples on days 3–4.

Table 1   *T_2 relaxation times and diffusion coefficients for the features observed in the no-salt Mozzarella-type cheese on day 3*

	T_2 relaxation time (s)	Diffusion coefficient (m^2 s^{-1})	Logarithmic co-ordinates [$\log_{10}T_2$, $\log_{10}D$]
Peak 1	0.220	1.4×10^{-9}	[–0.66, 0.15]
Peak 2	0.034	3.4×10^{-10}	[–1.47, –0.47]
Peak 3	0.054	1.9×10^{-11}	[–1.27, –1.72]
Peak 4	0.350	1.2×10^{-11}	[–0.46, –1.92]

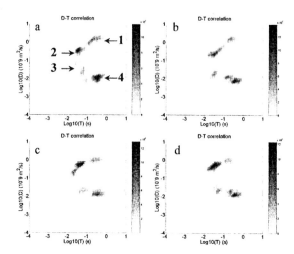

Figure 2   *T_2 versus D maps for water in Mozzarella-type cheese at 40°C on days 3 and 4: (a) no-salt, (b) low-salt, (c) standard-salt and (d) high-salt*

3.1 Peak 1

As the salt concentration increased, there was a decrease in the intensity of *Peak 1*, T_2 became shorter — 0.220 s (Figure 2a) for the no-salt sample compared with 0.170 s (Figure 2d) for the high-salt sample — and the diffusion coefficient became slower — 1.4 $\times 10^{-9}$ m^2 s^{-1} (Figure 2a) to 9.16 $\times 10^{-10}$ m^2 s^{-1} (Figure 2d).

By days 12–13, *Peak 1* appeared only in the no-salt sample. It was still present at day 29, but was absent by day 43. Age led not only to a decrease in the intensity of *Peak 1*, but also to decreases in T_2 and the diffusion coefficient to 0.135 s and 5.36 $\times 10^{-10}$ m^2 s^{-1} respectively (Figure 3c) by day 29.

The relaxation and diffusion characteristics of *Peak 1* were lower than those of free moving water (1-1.5 s and 2.3 $\times 10^{-9}$ m^{2}s^{-1} at 25°C [12]) as we would expect, however, both T_2 and D were relatively high when compared with the other features. We assign this feature as water in pools. Therefore, as the salt concentration increases, protein hydration also increases[12] and the signal from the water in pools diminishes.

This assignment is seemingly confirmed by a comparative study on butter at 40°C. Alastair MacGibbon (personal communication) showed, using pulsed gradient NMR methods, that the water in butter is contained in water droplets of various dimensions. Inspired by this, a two-dimensional correlation map for a sample of butter was obtained. It showed contamination from the broad oil peak at low diffusion coefficients (see Section 3.4 for details) and a second feature at 0.141 s and 1.4 $\times 10^{-9}$ m^2 s^{-1} [approximately –0.75, 0.25]. This compares very well with the age-dependent feature (*Peak 1*) we see in the Mozzarella-type cheese sample.

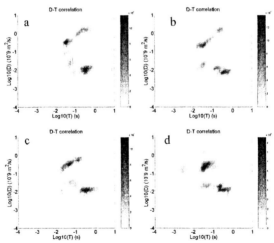

Figure 3   *T_2 versus D maps for water in no-salt Mozzarella-type cheese at 40°C on (a) day 3, (b) day 7, (c) day 29 and (d) day 42*

3.2 Peak 2

When the cheese was cooled to –10°C, only one feature remained; *Peak 2*. It is expected that water in close proximity to a large structure such as the protein will give a liquid-like NMR signal and therefore should still be observed below the freezing point of free water.[13,14] Low temperature studies suggest that *Peak 2* should be assigned as protein-associated water.

3.3   Peak 3

At the end of the study (days 114–115), the feature assigned to water in pools (*Peak 1*) was no longer evident; however, *Peaks 2, 3* and *4* still remained in all correlations. *Peak 3* was not observed at -10°C or in butter, ruling out both protein-associated water and pools of water.

Cheese-like gels with no fat (e.g. renneted casein with sodium hexametaphosphate) showed neither *Peak 3* nor *Peak 4* (an assignment for *Peak 4* is given in Section 3.4). However, most cheeses and cheese-like gels containing fat showed *Peak 3* and all showed *Peak 4*. These results give some indication that *Peak 3* was associated with the presence of fat.

Peak 1 is likely to arise from the water phase in an oil-in-water emulsion, found in the regions surrounding the milkfat agglomerations or strings of casein proteins formed during the stretching process. The pools of water are relatively large and the diffusion coefficient is high. Constriction of the size of these droplets will lead to restricted diffusion and a decrease in the diffusion coefficient.[15] In contrast to *Peak 1*, *Peak 3* was not age dependent, and indicated a remarkably slow diffusion in which the relaxation rate was slower than that ascribed to protein-bound water. These factors suggest that *Peak 3* may have been a result of small water droplets dispersed in an oil medium, water that was translationally highly immobilised and that could not easily become associated with or bound to the protein. These results indicate that a tentative assignment for *Peak 3* is a water-in-oil emulsion.

3.4   Peak 4

Peak 4 is assigned to a ghost image of the oil peak. It had similar relaxation and diffusion characteristics to those observed for the oil signal at 40°C. Comparison of the chemical spectra at 5°C and 40°C showed a broader oil peak that was shifted downfield as the fat melted. This caused the tail of the oil peak to leak magnetisation into the chemical shift region associated with the water signal. This led to the appearance of a ghost image when we integrated over this region to obtain the two-dimensional correlation map for water (Figure 4).

This feature was shown to be associated with oil protons using a chemical selective experiment that removed the majority of the oil signal. The results for the standard-salt Mozzarella-type cheese at 40°C using the chemical selective sequence showed reduced intensity of the peak assigned to the oil ghost.

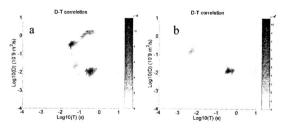

Figure 4   *T_2 versus D maps for no-salt Mozzarella-type cheese at 40°C on day 3: (a) water and (b) oil*

4 DISCUSSION

The present study showed that the correlation of diffusion coefficients and relaxation times could differentiate three distinct states or environments of the water in a group of Mozzarella cheeses. The major state was one characterised by a low diffusion coefficient and a rapid relaxation rate (*Peak 2*, Figure 2), which suggested that this water was both rotationally and translationally hindered. We assign this signal to water that is bound to the protein. The second type of water had a larger diffusion coefficient and a slower relaxation rate (*Peak 1*, Figure 2). These characteristics, which were similar to those of free water, suggested that this water was in pools and did not exchange readily with the protein-associated water. These two environments for the water correspond to the types of water identified by Kuo et al.[4], and called T_{21} and T_{22}, by a relaxation rate study at an NMR frequency of 400 MHz. Kuo et al. also found that both types of water were present in pasta filata (stretched curd) cheese but only the fast relaxation rate water T_{21} was present in the non-pasta filata Mozzarella cheese. They also noted that the T_{22} signal from the pasta filata cheese diminished with ripening time and interpreted this as resulting from the absorption of this type of water into the protein matrix (and thus becoming protein-associated water).

Guo and Kindstedt found that, in 2-day-old Mozzarella, a volume of aqueous liquid equivalent to one-third of the total cheese moisture content was expressed using centrifugation techniques.[16] By 2–3 weeks, no liquid was obtained. These results were similar to those we observed non-destructively using NMR. They found that little or no serum was expressed when the experiments were performed below 20°C. These results draw a parallel with our findings, in that the signal for pools of water was observed only in the experiments carried out at 40°C, not at 5°C. Perhaps our experiments could be optimised by reducing the quantity of melted fat and thus the effects of the oil ghost by carrying out the experiments at 25°C, as was done by Guo and Kindstedt.[16]

Our study also compares well with the results of Pastorino et al.[17] They found the influence of salt concentration to be most prevalent between 0 and 0.5%. We showed that only the no-salt sample (0.09%) showed any significant free water component beyond the first week of the study.

There was also a small discrete signal from water with a slower relaxation rate and a more limited diffusion (*Peak 3*, Figure 2). This low-intensity signal could have arisen from water encased within a fatty environment, although there are other possibilities (e.g. aqueous contents of bacterial cells). This signal had similar relaxation characteristics to *Peak 2* (protein-associated water) and would not be observed using relaxation alone, as was used by Kuo et al.[4]

Thus, this study not only supports the earlier NMR and centrifugable water data of Kuo et al.[4] and others[2,3,16], by showing that the water initially in pools is gradually absorbed into the protein matrix, but also extends upon these earlier NMR investigations by correlating diffusion and relaxation data and revealing another water population in pasta filata Mozzarella. Our study also combines the effects of both age and salt concentration on the sample.

5 CONCLUSIONS

The results of this NMR study show the value of the two-dimensional Laplace inversion correlation maps in the separation and assignment of water populations in Mozzarella-type cheese. As hypothesised, both age and salt concentration have an influence on the uptake of free water by the protein matrix.

It was possible to differentiate three types of water: protein-associated water, larger transient pools and small pools that were tentatively assigned as water-in-oil emulsions. The present results highlight the advantages of NMR in being able to observe water without the need to destroy the sample and introduce a visually attractive and immediately comprehensible method of distinguishing different water populations in many heterogeneous systems.

Acknowledgements

We thank Allan Main and Craig Honoré for initiating and supporting this work and Vaughan Crow, Keith Johnston and the Analytical Services Group for their help with the formulation and analysis of the cheeses. We also thank Mike Boland and Alastair MacGibbon for valuable discussions, and we gratefully acknowledge the financial support from the New Zealand Foundation for Research, Science and Technology (Contract DRIX0201). One of the authors (PLH) would like to thank Fonterra – Palmerston North for major funding.

References

1   C.J. Oberg, W.R. McManus and D.J. McMahon, *Food Struct.*, 1993, **12**, 251.
2   D.J. McMahon, C.J. Oberg and W.R. McManus, *Aust. J. Dairy Technol.*, 1993, **48**, 99.
3   P.S. Kindstedt, L.J. Kiely and J.A. Gilmore, *J. Dairy Sci.*, 1992, **75**, 2913.
4   M.I. Kuo, S. Gunasekaran, M. Johnson and C. Chen, *J. Dairy Sci.*, 2001, **84**, 1950.
5   P.T. Callaghan, S. Godefroy and B.N. Ryland, *Magn. Reson. Imaging*, 2003, **21**, 243.
6   S. Godefroy, L.K. Creamer, P.T. Watkinson and P.T. Callaghan, 'The use of 2D Laplace Inversion in Food Material', in *Magnetic Resonance in Food Science - Latest Developments*, eds., P.S. Belton, et al., The Royal Society of Chemistry, Cambridge, 2003, pp. 85-92.
7   T.P. Guinee, 'Growth/Survival of Bacteria in Cheese', in *Cheese: Chemistry, Physics and Microbiology*, ed. P.F. Fox, P.L.H. McSweeny, T.M. Cogan and T.P. Guinee, Chapman and Hall, London, 2003.
8   P.T. Callaghan, *Principles in Nuclear Magnetic Resonance Microscopy*, Oxford University Press, Oxford, 1991.
9   S.W. Provencher, *Comput. Phys. Commun.*, 1982, **27**, 229.
10   Y.-Q. Song, L. Venkataramanan, M.D. Hurlimann, M. Flaum, P. Frulla and C. Straley, *J. Magn. Reson.*, 2002, **154**, 261.
11   C.L. Lawson and R.J. Hanson, *Solving Least Squares Problems*, Prentice-Hall, Englewood Cliffs, New Jersey, 1974.
12. Holz, M., S.R. Heil, and A. Sacco, *Physical Chemistry Chemical Physics*, 2000. **2**, 4740.
13   P.S. Belton, R.R. Jackson and K.J. Packer, *Biochim. Biophys. Acta*, 1972, **286**, 16.
14   H.K. Lueng and M.P. Steinberg, *J. Food Sci.*, 1976, **44**, 1212.
15   P.T. Callaghan, K.W. Jolley and R.S. Humphrey, *J. Colloid Interface Sci.*, 1983, **93**, 521.
16   M.R. Guo and P.S. Kindstedt, *J. Dairy Sci.*, 1995, **78**, 2099.
17   A.J. Pastorino, C.L. Hansen and D.J. McMahon, *J. Dairy Sci.*, 2003, **86**, 60.

Subject Index

Printed in the United Kingdom
by Lightning Source UK Ltd.
109108UKS00001B/79-81